Structural and Chemical Characterization of Metals, Alloys and Compounds—2012

MATERIALS RESEARCH SOCIETY
SYMPOSIUM PROCEEDINGS VOLUME 1481

Structural and Chemical Characterization of Metals, Alloys and Compounds—2012

Symposium held August 12–17, 2012, Cancún, México

EDITORS

Dr. Ramiro Pérez Campos

Centro de Física Aplicada y Tecnología Avanzada, UNAM
Querétaro, México

Dr. Antonio Contreras Cuevas

Instituto Mexicano del Petróleo
San Bartolo Atepehuacan, México

Dr. Rodrigo A. Esparza Muñoz

Centro de Física Aplicada y Tecnología Avanzada,
UNAM
Querétaro, México

Materials Research Society
Warrendale, Pennsylvania

CAMBRIDGE
UNIVERSITY PRESS

CAMBRIDGE
UNIVERSITY PRESS

Shaftesbury Road, Cambridge CB2 8EA, United Kingdom

One Liberty Plaza, 20th Floor, New York, NY 10006, USA

477 Williamstown Road, Port Melbourne, VIC 3207, Australia

314–321, 3rd Floor, Plot 3, Splendor Forum, Jasola District Centre, New Delhi – 110025, India

103 Penang Road, #05–06/07, Visioncrest Commercial, Singapore 238467

Cambridge University Press is part of Cambridge University Press & Assessment, a department of the University of Cambridge.

We share the University's mission to contribute to society through the pursuit of education, learning and research at the highest international levels of excellence.

www.cambridge.org
Information on this title: www.cambridge.org/9781605114583

© Materials Research Society 2013

First published 2013

CODEN: MRSPDH

A catalogue record for this publication is available from the British Library

ISBN   978-1-605-11458-3   Hardback

CONTENTS

MICROSTRUCTURAL CHARACTERIZATION OF NANOSTRUCTURED MATERIALS

PREFACE

The XXI International Materials Research Congress was held in Cancun México from 12 to 17 August 2012. It was organized by Mexican Materials Society (SMM). About 1300 specialized scientists from more than 40 countries participated in the 28 different symposium, workshops, plenary lectures and tutorial courses. The 28 symposia that comprise the technical program of IMRC 2012 are grouped in several clusters, namely: Nanoscience and Nanotechnology, Materials Characterization, Materials for Energy Production, Biomaterials, Polymers, Electronic and Photonic Materials, Fundamentals Materials Science and General (Strategy for academy-industry relationship).

This Materials Research Society Proceedings contains papers presented at the Symposium 2D "Structural and Chemical Characterization of Metals, Alloys and Compounds" of the XXI International Materials Research Congress. This event is intended to be a forum for the dissemination of research results on materials research. The participants and the organizers have found this event very successful due to the high quality and novelty of the scientific results presented. Among the important achievements of the symposium were the new personal contacts between the scientists, for the creation of multinational thematic and research networks, as well as promoting contacts for future collaboration.

This special issue covers several aspects of the structural and chemical characterization of the materials in the following areas: metals, alloys, steels, composites, polymeric compounds, welding, nanomaterials, and surface coatings, among others. They are amorphous, crystalline, powders, coatings, fibers, thin films, etc., which were prepared with different techniques. The structural characterization techniques included: scanning electron microscopy (SEM), X-ray diffraction (XRD), transmission electron microscopy (TEM), RAMAN spectroscopy, optical microscopy (OM), Fourier transform infrared spectroscopy (FTIR), differential thermal analysis (DTA), differential scanning calorimetry (DSC), thermogravimetry analysis (TGA), thermo luminescence (TL), laser emission, etc. Theoretical models from these properties are included too.

The scientific program of symposium 2D included 67 oral and 146 poster presentations. In addition, invited talks were focused on different topics like X-ray diffraction, characterization of coatings and characterization of nanostructured materials. The special issue contains 16 papers based on contributions presented on the symposium. All manuscripts included in this special issue have been accepted after peer review.

We would like to express our deep acknowledgement to the Mexican Materials Society Advisory Committee, as well as sincere thanks to the reviewers for their valuable assistance and help in the review process. We also would like to thank the Mexican Materials Society (SMM), National University of México (UNAM) and Mexican Petroleum Institute (IMP) for the support in organization of the symposium.

Dr. Ramiro Pérez Campos
Dr. Antonio Contreras Cuevas
Dr. Rodrigo A. Esparza Muñoz

Editors

ACKNOWLEDGMENTS

We would like to thank the members of MRS-México advisory committee, as well as the reviewers for their valuable comments, which have certainly helped to improve the quality of the manuscripts. We also wish to thank the Mexican Materials Research Society, Universidad Nacional Autonoma de México (UNAM) and Mexican Petroleum Institute (IMP) for their support in organizing the symposium 2D "Structural and chemical characterization of metals, alloys and compounds".

Additionally, we would like to thank all those who have worked to make this congress an exciting and fruitful meeting, meeting chairs, symposia organizers, IMRC staff, MRS staff, editors, management committee, advisory committee, and Materials Research Society of México.

MATERIALS RESEARCH SOCIETY SYMPOSIUM PROCEEDINGS

MATERIALS RESEARCH SOCIETY SYMPOSIUM PROCEEDINGS

MATERIALS RESEARCH SOCIETY SYMPOSIUM PROCEEDINGS

Volume 1534E — Low-Dimensional Semiconductor Structures, 2012, T. Torchyn, Y. Vorobie, Z. Horvath,
ISBN 978-1-60511-511-5

Prior Materials Research Society Symposium Proceedings available by contacting Materials Research Society

Characterization of Materials by X-ray Diffraction (XRD), Scanning Electron Microscopy (SEM) and Differential Scanning Calorimetry (DSC)

Mater. Res. Soc. Symp. Proc. Vol. 1481 © 2012 Materials Research Society
DOI: 10.1557/opl.2012.1626

Synthesis and Determination of Thermodynamical Properties of the Compounds of the system Ca-Mg-Bi

C. Ramírez[1], J. A. Romero[1], A. Hernández[1], F. Pérez[2]
[1] ESIQIE-IPN, Z.C. 07738 Tel. 57296000 ext. 55270, México, D.F.
[2] University of the State of Hidalgo, Tel. 01-771-717-2000, ext. 2271, México.
E-mail: cony_ramirez_5@hotmail.com

ABSTRACT

Bismuth is an element obtained as a sub-product in lead production; Mexico occupies the second position in the world in production of this element. Bismuth is used as iron, aluminum and copper alloying, in the pharmaceutical industry, in the cosmetics industry, etc. Bismuth is separated from lead by the Kroll-Betterton Process in which a Ca-Mg alloy is added to the melting lead to form the intermetallic compounds Ca_3Bi_2, Mg_3Bi_2 and Ca_2MgBi_2 which float to the surface of the bath. Unfortunately, there is little thermodynamical information of the compounds of the system Ca-Mg-Bi which can be used to study and optimize the Kroll-Betterton process in a theoretical way. In this work there were synthesized the compounds Ca_3Bi_2, Mg_3Bi_2 and $CaMg_2Bi_2$ using powders of pure elements (Ca, Mg and Bi) in the required amounts and melted under an inert atmosphere. After synthesis, the samples were characterized by X-Ray Diffraction to ensure the formation of the desire compounds. Later, calorimetric technique was used to determine the thermodynamical properties of the compounds. The results obtained by X-Ray Diffraction show the formation of Mg_3Bi_2 and Mg_2CaBi_2 species; however, there is no crystallographic information of the compound Ca_3Bi_2. The heating curves obtained by calorimetry show endothermic peaks, due to the presence of phases changes as is indicated in the Ca-Bi, Mg-Bi and Ca-Mg-Bi phases diagrams.

Keywords: Bi, Mg, Ca, calorimetry, phase equilibria.

INTRODUCTION

Bismuth is mainly obtained as a sub-product of the refining of lead. This is because the lead and bismuth are elements with similar properties so that mineral deposits are associated with bismuth and lead. The common method for separating bismuth from lead is called "Kroll-Betterton Process" which uses calcium and/or magnesium to remove bismuth from lead bath. Kroll-Betterton Process consists in adding a calcium-magnesium alloy to form intermetallic compounds Ca-Mg-Bi of high melting point and lower density than lead. The standard specification of many countries stipulate a maximum of 0.005% bismuth in the metallic lead, nowadays, this is the most widely used process to recover bismuth from lead. Figure 1 shows the phase diagram Ca-Bi[1] where the compound Ca_3Bi_2 is formed with a composition of 60% at. Ca and 40% at. Bi. Figure 2 shows the phase diagram Mg-Bi[2] with the compound Mg_3Bi_2 with 60% at. Mg and 40% at Bi, this compound exhibits an allotropic transformation above 700 °C. Figure 3 shows the phase diagram Ca-Mg-Bi[3] where the ternary compound $CaMg_2Bi_2$ is formed with a composition of 20% at.Ca, 40% at. Bi and 40% at. Mg.

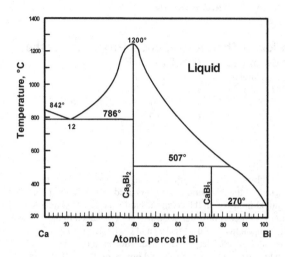

Figure 1. Phases diagrams of the system Ca-Bi [1].

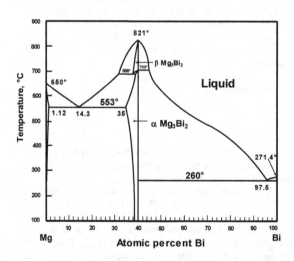

Figure 2. Phases diagrams of the system Mg-Bi [2].

4

Paliwal and Jung [4] determined the short range ordering behavior of liquid Mg-Bi solution and thermodynamic modeling of the Mg-Bi binary system had been optimized to obtain one set of model parameters for the Gibbs energies of the liquid and all solid phases as functions of composition and temperature. Kim[5] studied the thermodynamic properties of Ca-Bi alloys by electromotive force (emf) measurements at ambient pressure as a function of temperature between 723 K and 1173 K using a Ca(s)|CaF$_2$|Ca(inBi) cell for twenty different Ca-Bi alloys spanning the entire range of composition from $X_{Ca} = 0$ to 1. Kumar[6] studied the large deviation from the ideal mixture and the concentration dependent asymmetry in the thermodynamic properties of Mg-Bi alloy.

Invariant Reactions

1000°C e_1: Líq = Ca$_3$Bi$_2$ + CaMg$_2$Bi$_2$
712°C e_2: Líq = Ca$_3$Bi$_2$ + CaMg$_2$
640°C e_3: Líq = Ca$_3$Bi$_2$ + (Mg)
640°C e_4: Líq = CaMg$_2$Bi$_2$ + (Mg)
270°C e_5: Líq = CaMg$_2$Bi$_2$ + (Bi)
850°C p: Líq + CaMg$_2$Bi$_2$ = Mg$_3$Bi$_2$
630°C E_1: Líq = CaMg$_2$Bi$_2$ + Ca$_3$Bi$_2$ + (Mg)
551°C E_2: Líq = CaMg$_2$Bi$_2$ + Mg$_3$Bi$_2$ + (Mg)
516°C E_3: Líq = Ca$_3$Bi$_2$ + CaMg$_2$ + (Mg)
445°C E_4: Líq = Ca$_3$Bi$_2$ + Ca(HT) + CaMg$_2$
265°C E_5: Líq = (Bi) + CaBi$_3$ + CaMg$_2$Bi$_2$
258°C E_6: Líq = (Bi) + Mg$_3$Bi$_2$ + CaMg$_2$Bi$_2$
490°C U: Líq + Ca$_3$Bi$_2$ = CaBi$_3$ + CaMg$_2$Bi$_2$

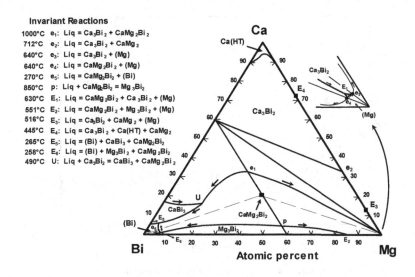

Figure 3. Phases diagrams of the system Ca-Mg-Bi[3].

EXPERIMENTAL PROCEDURE

For the experimental procedure first of all synthesis is defined with high purity powders of Calcium, Magnesium and Bismuth to form the intermetallic compounds Ca$_3$Bi$_2$, Mg$_3$Bi$_2$ and CaMg$_2$Bi$_2$ in the required proportions and melted at a temperature of 650° C under inert atmosphere during long time, the next step is the characterization by X-Ray Diffraction to confirm the formation of the intermetallic compounds, later is the determination of thermodynamic properties by calorimetry to determine the phase changes, invariant reactions and thermodynamic properties of the compounds.

5

RESULTS AND DISCUSSION

X-Ray Diffraction

Figures 4 to 6 show the results of X-Ray Diffraction of the systems Ca-Bi, Mg-Bi and Ca-Mg-Bi using the following conditions: voltage of 30 kV, current of 25 mA, scan speed 2 deg/min and a Cu Kα radiation = 1.5406. Figure 4 shows the X-Ray pattern of the sample Ca-Bi, where $Ca_{11}Bi_{10}$ and Ca_5Bi_3 are seen as a predominant species. It is also observed the presence of oxidized species and metallic bismuth. The compound Ca_3Bi_2 could not be detected by this technique since there is no crystallographic information of this compound.

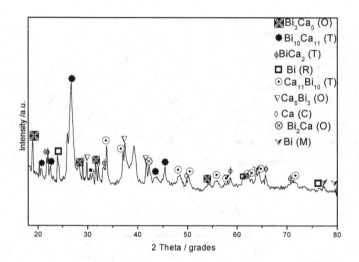

Figure 4. X-Ray pattern of the Ca-Bi system.

Figure 5 shows the X-Ray pattern of the Mg-Bi system which shows the presence of the compound Mg_3Bi_2 and Bi, possibly due to an excess of this element in the preparation of the compound. Figure 6 shows the X-Ray pattern of the Ca- Mg-Bi system where it is observed the formation of the ternary compound $CaMg_2Bi_2$ and Bi, possibly there was not a perfect homogenization of the powders and some amount of bismuth did not react.

6

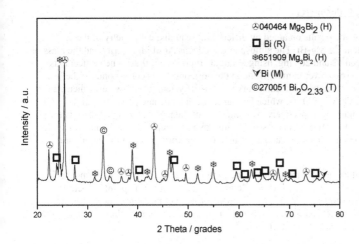

Figure 5. X-Ray pattern of the Mg-Bi system.

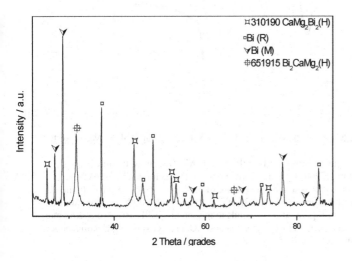

Figure 6. X-Ray pattern of the Ca-Mg-Bi system.

7

Differential Scanning Calorimetry

Figures 7 to 10 show the results obtained by differential scanning calorimetry of the compounds of the system Ca-Mg-Bi. The atmosphere used was Ar of high purity and the mass of the samples was between 12 and 264 mg, using alumina crucibles of 70 μL. The vertical axis shows the values of electric power in mW, while in the horizontal axis corresponds to the temperature (°C). Figure 8 shows the heating curve of Ca-Bi system where two endothermic peaks are obtained at 270° and 500° C, which belong to the eutectic and peritectic points of Ca-Bi binary diagram of Figure 1, respectively. The intensity of the peaks shows that there is little amount of free bismuth. However it is necessary to improve the preparation technique to avoid the formation of other chemical species and determine only the properties of the compound Ca_3Bi_2.

The heating curve of Figure 8 corresponds to the sample Mg_3Bi_2 and an endothermic peak is observed at a temperature of 282° C, which may correspond to the melting point of metallic bismuth. This curve also shows a peak at 395° C, however it is not possible to determine the transformation to which corresponds as there is not an invariant reaction or phase change at that temperature in the Mg-Bi phase diagram (Figure 2).

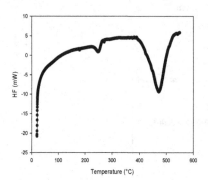 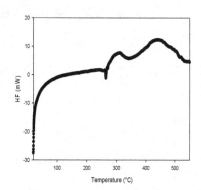

Figure 7. Heating curve of the Ca-Bi system.   **Figure 8.** Heating curve of the Mg-Bi system.

The heat capacity of Figure 9 corresponds to the sample Mg_3Bi_2, an exothermic peak is observed at the temperature of 260° C, which corresponds to the melting point of metallic bismuth. This graphic shows a peak at the temperature of 353° C which corresponds to the formation of the intermetallic compound αMg_3Bi_2 if it is related with the phases diagrams of figure 2. The heat capacity of Figure 10 corresponds to the sample $CaMg_2Bi_2$, two exothermic peaks are observed at the temperatures of 246° C and 446° C, respectively, which correspond to the formation of the invariant reactions in Figure 3.

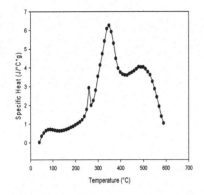

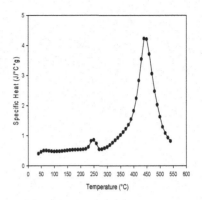

Figure 9. Specific heat versus temperature of Mg_3Bi_2 compound.

Figure 10. Specific heat versus temperature of $CaMg_2Bi_2$ compound.

CONCLUSIONS

This work deals with the preparation of compounds of the system Ca-Mg-Bi (Ca_3Bi_2, Mg_3Bi_2 and $CaMg_2Bi_2$) as well as their characterization. The main conclusions are:

- The intermetallic compounds Mg_3Bi_2 and $CaMg_2Bi_2$ were synthesized and analyzed showing an excess of Bi.
- The intermetallic compound Ca_3Bi_2 was not identified by X-Ray Diffraction due to the lack of crystallographic information of this compound.
- The heating curves obtained by calorimetry enables us to detect the phase transformations and associate them with the phase diagrams.
- We are still working with calorimetry technique to obtain the thermodynamical properties of the compounds of the Ca-Mg-Bi system.
- The curves of specific Heat-T were estimated for the Mg_3Bi_2 and $CaMg_2Bi_2$ compounds.

REFERENCES

1. *Handbook of Ternary Alloy Phase Diagrams*, ASM International, (1995).
2. T. B. Massalski, *"Binary Alloy Phase Diagrams"*, 2nd. Edition, ASM Int, Materials Park, Oh. (1990).
3. H. J. Okamoto, *"Phase Equilibria"*, 13(6), 649-679, (1992).
4. M. Paliwal, *"Thermodynamic modeling of the Mg-Bi and Mg-Sb binary systems and short-range-ordering behavior of the liquid solutions"*, CALPHAD: Computer Coupling of Phase Diagrams and Thermochemistry **33**, 744-754, (2009).
5. H. Kim, Electrochimica Acta **60**, 154-162 (2012).
6. A. Kumar, Advanced Materials Letters, 1-12, (2012).

Mater. Res. Soc. Symp. Proc. Vol. 1481 © 2012 Materials Research Society
DOI: 10.1557/opl.2012.1627

Study of the Synthesis of Mullite From Kaolin-α-Al$_2$O$_3$ and Kaolin-Al(NO$_3$)$_3$

E. M. Lozada, O. Alanís, F. Legorreta and L. E. Hernández

Centro de Investigaciones en Ciencias de la Tierra y Materiales, Universidad Autónoma del Estado de Hidalgo, Carr. Pachuca-Tulancingo km 4.5, C.P. 42184, Mineral de la Reforma, Hidalgo, México. E-mail: marle_loas@yahoo.com

ABSTRACT

The synthesis of mullite from kaolin clay and two precursors of aluminum: α-Al$_2$O$_3$ and Al(NO$_3$)$_3$ was investigated. In order to study the temperature effect, the system kaolin-α-Al$_2$O$_3$ was calcined in air in a range of 1200 to 1500°C, for 2 h. For the system kaolin-Al(NO$_3$)$_3$, the combustion method was employed, using urea as fuel, and calcined in air at 1500°C for 2 h. The products were characterized by X-ray diffraction, scanning electronic microscopy (SEM), energy dispersive spectroscopy and particle size analysis in order to analyze and compare their morphology and structure. The crystallographic study revealed an incomplete reaction between the kaolin and the α-Al$_2$O$_3$. Nevertheless, in the system kaolin-Al(NO$_3$)$_3$, it was obtained mullite with high purity and trace amounts of cristobalite.

Keywords: X-ray diffraction, structural, scanning electron microscopy (SEM), Al.

INTRODUCTION

Mullite (3Al$_2$O$_3$2SiO$_2$) is a refractory material which has been investigated by decades due to its high thermal stability, low density, high mechanical strength and good chemical stability [1-4]. In addition, the ability to operate at high temperature and in corrosive environments, allows mullite to be used in a variety of composites [5], as well as in advanced structural and functional ceramics [6]. Nevertheless, mullite is rarely in nature, as it requires high temperatures and low pressures [7]. Therefore, several methods have been proposed in order to obtain crystalline mullite from different starting materials [8-10].

Kaolinite (Al$_2$O$_3$2SiO$_2$H$_2$O), the major kaolin component, is the most common precursor for the synthesis of mullite [7-13] due to its potential and economic application. Nevertheless, the amount of SiO$_2$ in kaolinite is higher than in mullite; the excess SiO$_2$ together with the impurities in kaolinite forms a glassy phase and cristobalite to accompany the formation of mullite at temperatures higher than 1000°C [13]. The presence of a large amount of glass phase is thus detrimental to the mechanical properties of the mullite prepared from kaolinite [7]. In order to consume the SiO$_2$ in the glassy phase, Al$_2$O$_3$ is generally added so that the reaction product of SiO$_2$ and Al$_2$O$_3$ is also a mullite phase.

Solid state reaction is a useful method for introducing the Al$_2$O$_3$ into the kaolinite lattice which consists in the grinding of solids in a medium either dried or wet and their subsequent heat treatment at high temperatures. Parameters of synthesis such as calcination temperature, composition and texture of precursors as same as nature of the solvent, determine the final properties of the product [7, 14].

Furthermore, the implementation of a combustion reaction may improve the synthesis of mullite due to the addition of heat may favor the interaction between the SiO_2 and Al_2O_3 so that the interdiffusion rates of Si^{4+} and Al^{3+} within the mullite lattice may increase since it is typically slow thus restricts its formation [15]. In order to test this hypothesis, it was carried out the synthesis of mullite by using two methods: solid state reaction (SSR) and combustion-solid state reaction (CSSR) from kaolin extracted from Hidalgo, Mex., which is characterized by containing a significant proportion of quartz and some iron and titanium impurities.

EXPERIMENTAL PROCEDURE

Raw material

Kaolin clay (Molinos y Moliendas de Pachuca, S.A. de C.V.), which is extracted from the Tenango Dorian region in Hidalgo State, from Mexico, was used as starting material in the two methods of synthesis. It was processed mechanically using a roll crusher brand Quinn Process Equipment Company and a screen No. 400, in order to get a particle size mineral less than 37 μm. For removing SiO_2 impurities, a kaolin suspension was prepared with 45% w/w of solids and 1 kg/ton of sodium hexametaphosfate (Aldrich, 96%) as dispersant, milling at 300 rpm by 30 min. Then, deionizated water was added to the suspension in order to dilute at 20% w/w and was stirred vigorously. After 30 min, two phases were formed: suspended slurry, which was called dispersed kaolin, and the precipitated one, called assented kaolin.

Kaolin-α-Al₂O₃ system

5 g of kaolin clay and α-Al_2O_3 (Aldrich, 99.2%), with a molar fraction, x = n(α-Al_2O_3)/n($Al_2O_3 2SiO_2$) = 0.5, 1 y 2, were added together with 20 mL of ethanol (Aldrich, 99%) to a miller with zirconium balls and mixed at 3600 rpm during 2 h. The product was dried at 100°C and calcined at a temperature range of 1200 to 1500°C.

Kaolin-Al(NO₃)₃ system

5 g of kaolin clay was suspended in a solution with $Al(NO_3)_3$ (Aldrich, 99%) using a molar ratio n(α-Al_2O_3)/n($Al_2O_3 2SiO_2$) = 0.5, and a equivalent of $CO(NH_2)_2$ (Aldrich, 98%) as fuel, varying the solvent amount from 5 to 20 mL. Subsequently, it was carried out combustion at 600°C. The product was calcined at 1500°C.

Characterization

The composition of the materials were investigated by X-Ray Diffraction (XRD) (Inel, model EQUINOX 2000) and energy dispersive x-ray spectroscopy (EDS) and its microstructure was observed by scanning electronic microscopy (SEM) (Jeol JSM 6300). The average particle size was determined by laser ray analysis (Beckman Coulter, model LS13320).

RESULTS AND DISCUSSION

Morphology and composition of kaolin

As shown in Figure 1a, XRD pattern of kaolin clay exhibits intense peaks attributed to crystalline kaolinite and quartz in less proportion. Peaks around 30, 35, 50 and 55° reflect the presence of iron and titanium oxides in trace amounts, which are responsible of the kaolin beige tone. These impurities were corroborated by EDS and gravimetric analysis (see Table 1). Fe^{3+} and Ti^{4+} ions may substitute the octahedral Al^{3+} ions into the mullite lattice at high temperatures, which affects its porosity. However, this phenomenon has been observed when iron and titanium contents of starting materials are higher than 10% w/w [7]. From this, one can ensure that both iron and titanium impurities present in the raw material do not represent an important factor for the resultant textural properties of mullite. On the other hand, Figure 1b shows SEM results which allow seeing some plane and hexagonal kaolin particles with average size between 2, 8 and 22 nm which were corroborated by particle size analysis. These typical forms constitute the texture of kaolin.

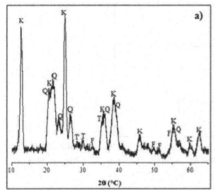 

Figure 1. (a) XRD kaolin pattern. K= kaolinite, Q= quartz, T=titanoferrum oxides and F= hematite. (b) SEM image of kaolin dried at 100°C.

Table 1. Chemical composition of kaolin.

Method	% w/w			
	Fe	**Ti**	**Al₂O₃**	**SiO₂**
EDS	0.54	0.45	38.62	57.38
GRAVIMETRY	0.48	-	-	-

Synthesis of mullite from kaolin-Al₂O₃ system

Since there was an Al_2O_3 defect, it was varied the molar fraction $n(\alpha\text{-}Al_2O_3)/n(Al_2O_32SiO_2)$ from 0.5 to 2 in order to study the chemical composition effect on the structure and morphology of mullite. As shown in Figure 2, no important differences were observed in XRD patterns, except in the diffraction peak around 22.5°, attributed to cristobalite formation which decreases as Al_2O_3 content increases. This suggests that a minor Al_2O_3 proportion reacts with SiO_2 to form mullite, but according to width of peaks around 26, 35, 37° and 53, as same as peaks at 44 and 58°, which indicate the presence of Al_2O_3 with corundum structure, there was no enough energy to favor a complete reaction between SiO_2 and Al_2O_3, spite of the exhaustive heat treatment at 1500°C.

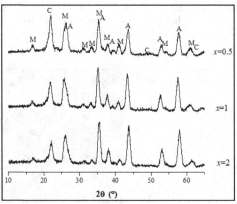

Figure 2. XRD patterns of mullite from kaolin-α-Al_2O_3 system with a molar fraction $x = n(\alpha\text{-}Al_2O_3)/n(Al_2O_32SiO_2)$ = 0.5, 1 and 2, calcined at 1500°C. M=mullite, A=alumina and C= cristobalite.

On the other hand, the temperature effect on the mullite synthesis was studied on the kaolin-α-Al_2O_3 system with a molar fraction $x = n(\alpha\text{-}Al_2O_3)/n(Al_2O_32SiO_2)$ = 0.5, varying the calcinations temperature since 1200 to 1500°C. As shown in Figure 3, XRD patterns evidence an improvement of mullite crystallinity since 1300°C.

14

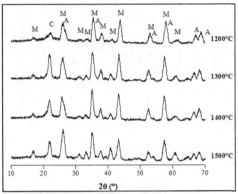

Figure 3. XRD patterns of mullite from kaolin-α-Al$_2$O$_3$ system with a molar fraction x = n(α-Al$_2$O$_3$)/n(Al$_2$O$_3$2SiO$_2$) = 0.5, calcined at different temperatures. M=mullite, A=alumina and C= cristobalite.

In addition, diffraction peak around 22.5°C associated to cristobalite, decreases slightly as the temperature increases, which stated a positive influence of a heat treatment at high temperature. Nevertheless, the presence of diffraction peaks attributed to corundum structure, bears out that the method of synthesis lacks of enough energy to favor the mullite formation.

Synthesis of mullite from kaolin-Al(NO$_3$)$_3$ system

In order to get a closed interaction between SiO$_2$ and Al$_2$O$_3$, it was carried out a combustion-solid state reaction from the kaolin-Al(NO$_3$)$_3$ system with a molar fraction x = n(α-Al$_2$O$_3$)/n(Al$_2$O$_3$2SiO$_2$) = 0.5, using urea as fuel and varying the solvent amount since 5 to 20 mL. According to the definition and intensity of the diffraction peaks attributed to mullite observed in Figure 4, it follows that the solvent amount plays an important role, because it facilitates an increased interaction between the particles of the reactants, so that the XRD pattern of mullite obtained using 20 mL of solvent shows more defined and intense peaks which reveal a major crystallinity. However, the presence of cristobalite, whose diffraction peaks are in proportion with mullite ones, suggests that quartz contained in the starting kaolin, which is converted to cristobalite around 1400°C, does not react, so that ultimately appears as a mullite impurity. From these results one can deduce that initial composition of raw material has a main influence on the mullite purity independently the employed method of synthesis.

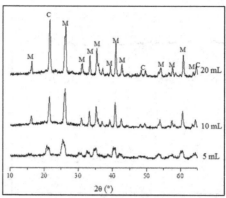

Figure 4. XRD patterns of mullite from kaolin-Al(NO$_3$)$_3$ system with a molar fraction x = n(α-Al$_2$O$_3$)/n(Al$_2$O$_3$2SiO$_2$) = 0.5, using different solvent amount and calcined at 1500°C. M=mullite and C= cristobalite.

Kaolin composition effect in the synthesis of mullite

As discussed earlier, kaolin clay extracted from the Tenango Dorian region in Hidalgo State, is characterized for containing large amount of quartz, which agglomerate and form big particles with diameter sizes higher than 10 μm. According to Legorreta, et al. [16], these particles can be separated by adding a dispersant to a kaolin suspension in a low proportion. In this case, sodium hexametaphosphate was used as dispersant whereby SiO$_2$ content in raw material was reduced significantly. Figure 5 shows the XRD patterns of mullite obtained from dispersed kaolin, which showed an important intensity decrease of the diffraction peaks around 22, 47 and 48°, which are attributed to cristobalite form.

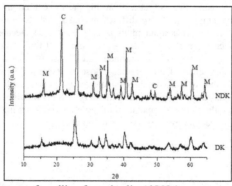

Figure 5. XRD patterns of mullite from kaolin-Al(NO$_3$)$_3$ system with a molar fraction x = n(α-Al$_2$O$_3$)/n(Al$_2$O$_3$2SiO$_2$) = 0.5, using not dispersed kaolin (NDK) and dispersed kaolin (DK) and calcined at 1500°C. M=mullite and C= cristobalite.

It was demonstrates that the formation of crystalline mullite with a high purity have an important dependency on the raw material composition, which, as the same time, can be modified by different pretreatments.

CONCLUSIONS

Two methods of synthesis of crystalline mullite were probed in order to study the effect of different parameters such as calcinations temperature, nature and quantity of solvent and precursors. According to XRD patterns, it was determined that as calcinations temperature as solvent quantity play an important role in the crystallinity of mullite, which improves visibly as these factors are increased. On the other hand, it was evidenced that CSSR method is an efficient and inexpensive way to prepare crystalline and pure mullite, despite the use of poor kaolin clay as raw material, since it favors a major interaction between the particles of the precursors. Therefore, it could be considered as an alternative method to improve structural properties of mullite. Finally, dispersed kaolin, which implied a less initial content of SiO_2, allowed obtaining mullite with a good purity. This suggests that modifying the composition of the starting materials, one can improve the yield of pure mullite despite of the formation of SiO_2 during its synthesis.

REFERENCES

1. I. H. Aksay, D. M. Dabbs and M. Sarikaya, *J. Amer. Ceram. Soc.* **74**, 2343 (1991).
2. B. R. Marple and D.J. Green, *J. Amer. Ceram. Soc.* **71**, C471 (1998).
3. B. Kanka and H. Schneider, *J. Mat. Sci.* 29, 1239 (1994).
4. M. D. Sacks, K. Wang, G. W. Scheiffele and N. Bozkurt, *J. Amer. Ceram. Soc.* **80**, 663 (1997).
5. B. J. Kanka and H. Schneider, *J. Europ. Ceram. Soc.* **20**, 619 (2005).
6. G. Celloti, I. Morettini and G. Ortelli, *Mater. Sci.* **18**, 1005 (1983).
7. H. Scheneider and S. Komarneni: *Mullite* (Wiley-VCH Verlag GmbH & Co. KGaA, Weinheim 2005).
8. C.Y. Chen, G.S. Lan and W.H. Tuan, *J. Eur. Ceram. Soc.* **20**, 2519 (2000).
9. T. Ebadzadeh, M. H. Sarrafi and E. Salahi, *Ceram. Int.* **35**, 3175 (2009).
10. B. Simendic and L. Radonjic, *J. Therm. Anal. Cal.* **56**, 199 (1999).
11. S. G. Dolgikh, A. K. Karklit, V. P. Migal and G. A. Karas, *Refractories.* **36**, 65 (1995).
12. K. Okada and N. Otsuka, *J. Am. Ceram. Soc.* **69**, 652 (1986).
13. J. A. Pask and A. P. Tomsia, *J. Am. Ceram. Soc.* **74**, 2367 (1991).
14. Y. Zhang, C. Xiao and J. Yang, *J. Sol-Gel Sci Technol.* **57**, 142 (2011).
15. W.E. Lee and W.M. Rainforth, *Ceramic Microstructures and Property Control by Processing* (Chapman & Hall, London, UK, 1994).
16. F. Legorreta-García, L. E. Hernández-Cruz, P. F. Mata-Muñoz: *Rev. Latinoam. Metal. Mat.* Accepted on October 2012.

Mater. Res. Soc. Symp. Proc. Vol. 1481 © 2012 Materials Research Society
DOI: 10.1557/opl.2012.1628

Particle Size Characterization of Commercial Raw Materials and Graphite nanoparticles of a Refractory Bricks Mix of the System Al$_2$O$_3$-SiC-C

A. M. Paniagua[*], J. Martinez, V. Mauro, E. Diaz

Escuela Superior de Física y Matemáticas Edificio 9, U.P. Adolfo López Mateos, Col. San Pedro Zacatenco, C.P. 07730 D. F., México.

[*] E-mail: ampani@esfm.ipn.mx.

ABSTRACT

In this work was studied the partial substitution in the design of a refractory mix of the fine crystalline graphite for prefabricate nanoparticles of the seam source of graphite, improving the refractory properties of the material, getting a better resistant to the chemical attack by the slag and steel liquid metal. The raw materials and nanoparticles of crystalline graphite were characterized by X-ray diffraction (XRD), Sherrer equation, and scanning electron microscopy (SEM). The nanoparticles size determines the crystalline of the graphite used in the mixes obtained after different steps of mechanic milling. The nanoparticles of materials were added to the mixes in different proportions. The commercial raw materials used for this investigation were: commercial silicon carbide high purity (97% SiC), calcined bauxite (85% Al$_2$O$_3$), alpha calcined alumina, and crystalline graphite (94% C). Additionally, six different sizes of graphite nanoparticles were selected. The particle size of the initial commercial graphite was 0.044mm and the final nanoparticles obtained in this investigation by mechanic milling was 18 nm. The measurement of the particle size of the nanoparticles was made by the Scherrer equation, XRD and SEM.

Keywords: nanostructure, X-Ray diffraction, crystalline, scanning electron microscopy (SEM).

INTRODUCTION

The extreme operative working conditions (high temperature and chemical attack) that is subjected the refractory lining in the steel furnaces production and, if we add to this critical stage, the direct contact between the refractory with the liquid slag and steel is required a refractory material with outstanding properties; the roll of the raw materials used in the design of a refractory mix for this application is very important, in the last 20 years have been performed some refractory brick mixes with a very good successful in the production of steel, some of the most important is the system Al$_2$O$_3$-SiC-C. The main raw materials for the formulation of a commercial refractory brick of this type are: Alumina (Al$_2$O$_3$), Silicon carbide (SiC) and Graphite (C), one of the main destructive mechanisms is the chemical corrosion on the refractory surface which is very severe, and their mechanical properties are affected in a negative and detrimental way.

The refractory materials are very necessaries in developed countries and emergent economies due the employment in critical conditions for industrial process as burning, fusion, calcinations, and clinkerization; developed in furnaces by endothermic and exothermic reactions with temperatures > 1400 C. These processes to the conditions are used as materials for refractory lining of melting pot, rotatory kilns, racks, boilers between others applications.

The industry of refractory materials have experiment a big evolution in the ultimate years as consequence of nanotechnology [1, 2] and requirements of consumer industry [3].

An alternative of the traditional systems of refractories are the bricks Al_2O_3-SiC-C, that principal characteristics is low thermal expansion, resistance to cracking, support repeat cycles of heating and cooling and corrosion resistance [4,5,6]. The superior mechanical properties oxide-C is related with inelastic deformation (flexibility) due to graphite presence [7, 8].

The objective of this work is the characterization of the crystalline nanoparticles graphite with different nanosizes by X-ray diffraction, Sherrer equation, and SEM, and morphology in their deposit on the mixes of refractory bricks Al_2O_3-SiC-C.

EXPERIMENTAL DETAILS

The refractory bricks of system Al_2O_3-SiC-C with nanoparticles of crystalline graphite were made with different sizes of raw materials with mesh 9-14. The dry mixes were bauxite (21%), graphite (10%), and Aluminate oxide (10%), Silicon Carbide (10%) and powder resin (4%) as binder mix. The preparation of graphite nanoparticles was made in dry by a mechanic milling with different milling times, in a planetary ball mill FRITSCH Pulverisette 5, with 230 rpm and agate balls and diameter of 2.0 cm.

The graphite particles were characterized by X-ray diffraction technique and Sherrer equation, in the diffractometer Brucker D8 Advanced with Cooper lamp and a CuKα (λ=1.54178 Å), voltage de 35 KV y I = 25 mA, ø =10 a 120°.

The diffractions were indexed using the data base ICDD PDF2 Release 2003 and the refinement of pattern by use of software Pick Fit, velocity of 2 °C/min. The identity peaks were used with Sherrer Equation for determinate the particle size. The morphology and size was too determinate by High resolution SEM FEI Sirion with resolution de 5nm. The raw materials were mixed with the nanoparticles of graphite, resin, water, pouring in a steel brick mould and pressed.

RESULTS AND DISCUSSION

X-ray Diffraction

Figure 1 shows the powder diffraction patterns obtained from raw materials. The X-ray Diffraction patterns of the materials for the system Al_2O_3-SiC-C reported α-Al_2O_3 with an orthorhombic structure and a red parameters a=4.38Å, b=8.31Å and c=8.94Å. The crystalline graphite phase structure is an orthorhombic structure and a red parameters of a=4.7Å, b=5.97 Å, c=4.4 Å, and Silicon Carbide phase with a rhombohedra structure and a red parameters of a=06.8Å, c=4.73 Å. The influence of Al_2O_3 consist in promote the SiC influencing in the sinters and to get formation of SiO_2.

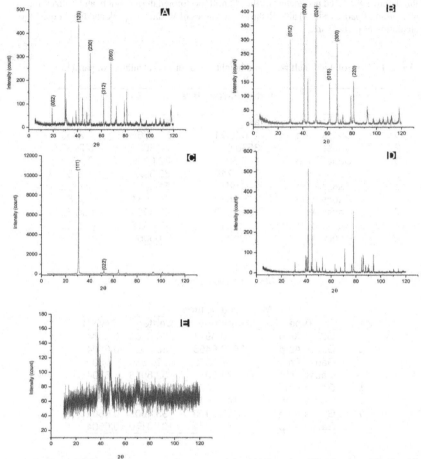

Figure 1. X-ray Diffractions patterns of raw materials: (A) Bauxite, (B) Alumina, (C) Graphite 2H, (D) Silicon carbide, (E) Slag.

Figure1 corresponding the characteristics peaks of bauxite, alumina, graphite, silicon carbide and slag which will be used in the corrosion of bricks with nanoparticles, this is amorphous with some copper peaks, the slag become of a furnace.

Table I shows the fitted parameters of the 2nd mill, performed during 148 h, and utilized for the calculation of particle size. Table II shows the values of Width half peak and height used for the calculation of particle size.

Table I. Value of width half peak, and height, used for the calculation of particle size.

		Fitted Parameters		
Peak	Type	a_0	a_1	a_2
1	Gauss Amp	170.3971	29.0763	0.2762
2	Gauss Amp	3530.645	30.8027	0.2762
3	Gauss Amp	147.5008	32.0089	0.2762
4	Gauss Amp	74.00710	32.8028	0.2762
5	Gauss Amp	140.6869	51.8660	0.2762
6	Gauss Amp	148.5513	64.2484	0.2762
7	Gauss Amp	106.9470	93.4005	0.2762
8	Gauss Amp	80.1981	101.3170	0.2762
B	Linear Bg	28.6635	0.0505	0.2762

Table II. Measured values.

	Measured Values			
Peak	Type	Amplitude	Center	FWHM
1	Gauss Amp	170.3971	29.0763	0.6504
2	Gauss Amp	3530.6453	30.8027	0.6504
3	Gauss Amp	147.5008	32.0089	0.6504
4	Gauss Amp	74.0071	32.8028	0.6504
5	Gauss Amp	140.6869	51.8660	0.6504
6	Gauss Amp	148.5513	64.2484	0.6504
7	Gauss Amp	106.9470	93.4005	0.6504
8	Gauss Amp	80.1981	101.3170	0.6504

Particle size determination by Sherrer Equation

Particle size of graphite nanoparticles was determined essentially by the Scherrer equation. Table II presents the values obtained from the powder diffraction pattern and utilized for calculating the particle size in milling. Equation (1) shows the Scherrer equation used for calculation of particle size of graphite nanoparticles.

$$t = \frac{0.9\lambda}{F\cos\theta} \qquad (1)$$

The influence that have the milling time in the particle size is the decreased of 0.044 mm to 18nm in 148 h as is shown in figure 2. The milling was carried out in dry. With the milling graphic is possible determinate the time necessary for get a minor particle size and the time. In addition, is possible know the equation for the curve too.

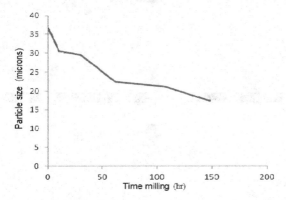

Figure 2. Time milling vs particle size of graphite.

Scanning Electron Microscopy

The morphology of milling graphite consist in particles size to 20μm-10nm (figure 3), in the micrograph can see particle laminate and agglomerates that are products of mechanic milling, with size varying in μm.

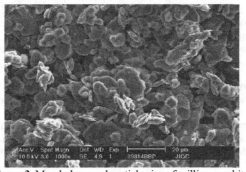

Figure 3. Morphology and particle size of milling graphite.

Figure 4 shows the morphology of green bricks of the system Al_2O_3-SiC-C. The influence of incorporate nanoparticles is to get more density, more corrosion resistance, to decrease the intergranular voids in the matrix of the system, and to increase mechanical properties in the bricks system. For better compaction, big and fine particles are needed. This avoids segregation, a low density and a system with poor mechanical properties, figure 4A.

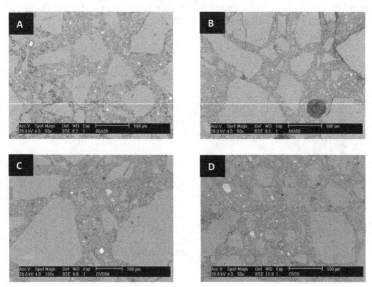

Figure 4. Morphology of green bricks system Al_2O_3-SiC-C.

Figure 4B is also appreciated a pore where there are particles precipitated with an approximate size of 1-5 microns. The Cracks are propagated by a "stop and go" mechanism: a crack is propagated at constant rate along grain boundaries with stops by readjustment of the direction of crack propagation at triple grain boundary junctions.

Figures 4C and D is observed a homogeneous matrix, where grain boundaries likewise refractory are delimited, and the presence of few particulates. In the matrix is observed reduced porosity unhydrated.

The graphite nanoparticles are deposited as filler of pores and voids [9-12], the porosity is decreased by influence of particle size, observed in matrix, Figure 4D. Cracks and micro cracks are also observed.

In the micrographs, micro cracks are observed unhydrated, as well as, areas for the binder. The apparent dependence grain size–wear rate in the materials is shown. The materials with finer microstructure have higher wear resistance and exhibit a monotonous relationship between wear rate and the grain size. Moreover, some authors observed the change of fracture mode in Al_2O_3– SiC composites, irrespective of the size and location of SiC particles, which they attributed to chemical, rather than a mechanical effect [12]. However, a meticulous image analysis shows that, upon decreasing the SiC content, amount of plastic deformation is enhanced and the reduction of both the intra- and inter-granular fracture are observed as the wear rate of the materials decreases. We attribute this effect to one competing mechanism: (I) grain boundary weakening caused by intergranular SiC particles, which gradually decreases with decreasing SiC content.

CONCLUSIONS

- The particle size decrease by mechanic milling getting more crystalline material with polymorphism that is the base for Al_2O_3-SiC-C. Knowing the mill is possible determinate the milling time and particle size.

- By Sherrer equation and scanning electron microscopy the nanoparticles size can be possible exactly determinate.

- The graphite particles and nanoparticles will be deposited between particles as filler of pores, voids, in the matrix, for better corrosion resistance, mechanical properties, and chemical attack.

REFERENCES

1. E. Karamian A. Monshi, A. Bataille, A. Zadhoush, *J. of the European Ceramic Society* 31,14, 2677–2685, (2011).

2. S. Tamura, T. Matsui, T. Ochiai, K. Goto; Technological philosophy and perspective of nanotech refractories. *Nippon Steel Technical Report* 98,18-28, (2008).

3. K. Sugita; Historical overview of refractory technology in the steel industry. *Nippon Steel Technical Report* 98, 8-17, (2008).

4. S. Camelli, M.J. Rimoldi, P. Marinelli, J.J. Mirabelli; Evaluation of different wear mechanism of Al_2O_3-SiC-C bricks used in torpedo ladle.7 th IAS *Ironmaking Conference. Campana, Buenos Aires, Argentina*, 192-201, (2009).

5. W. Schärfl, C.G. Aneziris, U., Klippel, V. Roungos; Properties and processing of shaped alumosilicate-carbon composites for application in shaft furnaces. *http://ekw-feuerfest.de/*

6. S. Ito, T. Inuzuka; Technical Development of Refractories for Steelmaking processes. *Nippon Steel Technical Report* (98), 63-69, (2008).

7. V. Miñoz, G.A. Rohr, A.G. Tomba-Martinez, A.L Cavalieri; Aspectos experimentales de la determinación de curvas esfuerzo-deformación a alta temperatura y en atmosfera controlada:

Refractarios Al_2O_3-MgO-C. *Boletín de la Sociedad Española de Cerámica y Vidrio* 50,3, 125-134, (2011).

8. M. Hassan-Amin, M. Amin-Ebrahimabadi, M. Reza-Rahimipour, *Journal of Nanomaterials*, 325674-325678, (2009).

9. G. Ruan, Y. Dong, Z. Zhang, S. Zhou, *Rare Metals*, 501-505, (2011).

10. C. Chen Feng, A. B. Bernard, L.E. William, *Journal of the American Ceramic Society*, 3177-3188, (1998).

11. S. Uchida, K. Ichikawa, K. Niihara, *Journal of the American Ceramic Society*, 2910-2916 (1998).

12. W.E. Lee, B.B. Argent, S. Zhang, *Journal of the American Ceramic Society*, 2911-2918, (2002).

Characterization of Materials by Differential Thermal Analysis (DTA), Thermogravimetry (TGA), Fourier Transform Infrared Spectroscopy (FTIR), Transmission Electron Microscopy (TEM) and RAMAN Spectroscopy

Mater. Res. Soc. Symp. Proc. Vol. 1481 © 2012 Materials Research Society
DOI: 10.1557/opl.2012.1629

Preparation of Lithium Aluminum Layered Double Hydroxide from Ammonium Dawsonite and Lithium Carbonate

C. A. Contreras Soto, E. Ramos-Ramírez, V. Reyes Zamudio and J. I. Macías.

Universidad de Guanajuato, Campus Guanajuato, División de Ciencias Naturales y Exactas, Departamento de Química, Noria alta s/n, Col. Noria alta, Guanajuato, Gto., México. Tel.: (+05) 473-73 20006 Ext. 6022.

E-mail: cesarcon@ugto.mx

ABSTRACT

Aluminum lithium hydroxide carbonate hydrate, also known as Al/Li layered double hydroxide or Al-Li hydrotalcite-like compound $[Al_2Li(OH)_6]_2CO_3 \cdot nH_2O$, was prepared by reaction of lithium carbonate with ammonium dawsonite $[NH_4Al(OH)_2CO_3]$. The reaction of ammonium dawsonite with a lithium carbonate sature solution at different temperatures and lithium carbonate concentrations was studied. The obtained solids were characterized by differential thermal analysis (DTA), thermogravimetry (TGA), Fourier transform infrared spectroscopy (FTIR), X-ray diffraction (XRD) and transmission electron microscopy (TEM). By this method, crystalline Li/Al LDH $[Al_2Li(OH)_6]_2CO_3 \cdot 3H_2O$ can be obtained at 60 °C and 4 h reaction time.

Keywords. Al, Li, chemical synthesis, thermogravimetric analysis (TGA), transmission electron microscopy (TEM).

INTRODUCTION

The terms layered double hydroxides (LDHs) or hydrotalcite-like compounds (HTLc) are used to designate synthetic or natural lamellar hydroxides with two or more kinds of metallic cations in the main layers and hydrated interlayer domains containing anionic species. LDHs may be synthesized with a wide range of compositions and a large number of materials with a wide variety of M(II)/M(III) cation combinations as well as M(I)/M(III) cation pairs (e.g. Li/Al) with different anions in the interlayer can be obtained [1]. Li/Al hydroxycarbonate hydrate $[Al_2Li(OH)_6]_2CO_3 \cdot nH_2O$, with laminar structure, belongs to the LDHs family, and is obtained by hydrolysis of Al^{3+} in the presence of lithium carbonate solutions [2]. LDHs possess very interesting properties with applications in catalysis, adsorption, pharmaceutics, and other areas [3-5]. Owing to its technical importance, several synthesis methods have been developed as for example coprecipitation, reconstruction, sol-gel technique, hydrothermal, microwave, and ultrasound treatments, anion-exchange reactions, among others [6].

In a previous work, Li/Al HLC was prepared by reaction of basic aluminum sulfate with lithium carbonate [7]. However, the sulfate anion present in the final aqueous solution must be eliminated from solution in order to the lithium carbonate solution can be used again in the process. In this work, ammonium dawsonite [$NH_4Al(OH)_2CO_3$] is proposed as a precursor of Li/Al LDH. The use of ammonium dawsonite as a precursor permits the mother liquid to be reused because of elimination of the ammonium ion from solution owing to the high solution pH attained at the end of the process.

EXPERIMENTAL

The ammonium dawsonite precursor (NH_4-Dw) was prepared by reacting basic aluminum sulfate with ammonium carbonate at 60 °C, for four hours as reported elsewhere [8]. The as obtained solid was separated by vacuum filtration and dried at 90 °C in an electric oven before to be analyzed.

The amount of lithium carbonate required to Li/Al LDH formation was determined as follows: 0.5 g of ammonium dawsonite was added to three Erlenmeyer flask, respectively. Then, 4.1, 4.9 and 6.6 mmol of Li_2CO_3 were added to the flasks, respectively, from a saturated solution of lithium carbonate (Baker) containing 12.12 g of Li_2CO_3 per liter. Water was added to each one of the flasks in order to attain 25 mL of liquid. The flasks were heated at 60 °C, for four hours in a heating bath in order to Li/Al LDH forms. These temperature and time were chosen because of these were the best experimental conditions for preparation of Li/Al LDH from basic aluminum sulfate, as reported in a previous work. After heating, the solids were separated from the liquid by vacuum filtration, washed with distilled water and dried in an oven at 90 °C before to be analyzed.

Once the concentration of lithium carbonate was determined, the effect of reaction time was investigated. In this case, the previously determined amount of lithium carbonate necessary for Li/Al LDH formation was added to three samples containing 0.5 g of ammonium dawsonite, respectively. The samples were heated at 60 °C for two, four and six hours, respectively. The solids were separated from the liquid by vacuum filtration, washed with distilled water and dried in an oven at 90 °C before to be analyzed. The flowchart of the synthesis process is presented in Figure 1.

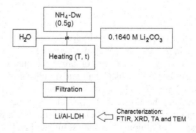

Figure 1. Flowchart of the synthesis process of Li/Al LDH.

The effect of heating temperature on Li/Al LDH formation was studied. Four samples containing 0.5 g of ammonium dawsonite and the previously determined amount of lithium carbonate were prepared. The samples were heated at 30, 60,70, 80 and 90 °C for four hours in a heating bath in order to Li/Al LDH forms. The solids were separated from the liquid by vacuum filtration, washed with distillated water and dried in an oven at 90 °C before to be analyzed.

The chemical structure of the as obtained solid was determined by Fourier Transform Infrared Spectroscopy (FTIR) in a Fourier transform Perkin-Elmer 1600 series spectrometer using the KBr technique. Each transparent pellet was made by pressing a mixture of 0.5 mg sample and 200 mg of spectra-pure KBr. The composition of the sample was determined by Thermal Analysis (TGA, DTG, and DTA) in a TA SDT Q600 thermal analyzer by heating around 0.3 mg of sample in a platinum crucible at a rate of 10 °C/min in a dynamic air atmosphere (100 cm^3/minute). The crystalline phases in the solid was identified by powder X-ray Diffraction spectroscopy (XRD) in an Siemens D-500 diffractometer coupled to a copper anode tube using monochromatized Cu Kα1 radiation. The acceleration voltage and the current flux were 30 kV and 20 mA, respectively. Samples for XRD were prepared as thin layers on the Plexiglassample holder.The size and morphology of the particles were studied by Transmission Electron Microscopy (TEM) using a Jeol JEM-100S. In this case, the sample powders was suspended in ethanol and ultrasound irradiated for five minutes in an ultrasonic cleaner (Branson 1510). Then, 7 μL of the suspension was measured and placed in a TEM grid prior to be analyzed.

RESULTS AND DISCUSSION

The precursor ammonium dawsonite was prepared and characterized by FTIR , XRD and TA. The analysis confirms the presence of well crystalized ammonium dawsonite. The thermal analysis traces (TGA and DTA) of ammonium dawsonite is shown in Figure 2. The chemical composition of this compound, calculated from thermal analysis data is: Al$_2$O$_3$, 36.6%; CO$_2$ + H$_2$O, 61.5%. The theoretical content of Al$_2$O$_3$ is 36.7%.

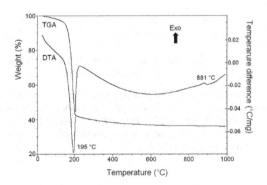

Figure 2. TGA and DTA curves of ammonium dawsonite.

Li/Al DLH preparation

In Figure 3, the FTIR spectra corresponding to the solids obtained after addition of 4.1, 4.9 and 6.6 mmol of Li_2CO_3 to 0.5 gram of ammonium dawsonite, respectively, are shown. All samples exhibit the strong band at 3448 cm^{-1} that can be assigned to stretching vibration of O-H groups in the hydroxide structure, as well as in water physically adsorbed. The wide band at 1615-1530 cm^{-1} can be attributed to O-H bending vibrational modes of molecular water [9]. The bands at 1366 cm^{-1} and 1006 cm^{-1} correspond to asymmetric stretching modes of CO_3^{2-} [10]. Finally, the absorption peaks at 731, 535 and 462 cm^{-1} can be attributed to Al-O stretching vibrations of octahedral AlO_6 [11]. The noticeable sharpness of the lattice absorption bands at 535 and 462 cm^{-1} indicates the octahedral cation ordering. All the absorption bands of Al-Li LDH sample coincide well with a Li/Al LDH obtained from aluminum-tri-(sec-butoxide) and lithium carbonate [11]. Because of the intensity and sharpness of the absorption peaks, the amount of lithium carbonate necessary to form Li/Al LDH is considered to be 6.6 mmol Li_2CO_3 per 0.5 g ammonium dawsonite.

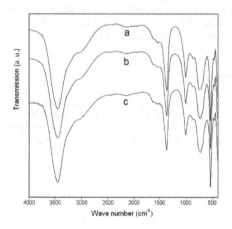

Figure3. FTIR spectra of the solids obtained after addition of: a) 4.1; b) 4.9; and c) 6.6 mmol Li_2CO_3 per 0.5 g ammonium dawsonite.

The FTIR spectra corresponding to the samples prepared after heating for two, four and six hours at 60 °C are shown in Figure 4. As can be seen, all spectra show the characteristic peaks of Li/Al LDH. However, the traces of the samples obtained after four and six hours heating time exhibits well defined absorption peaks. Therefore, four hours heating time is considered to be the adequate reaction time for crystallization of Li/Al LDH.

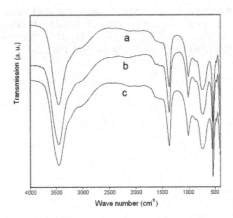

Figure 4. FTIR spectra of the solids obtained at 60 °C, after:a) 2 h;b) 4h; and c) 6 h.

The effect of temperature on the formation of Li/Al LDH was determined by FTIR. Li/Al LDH formed at temperatures between 60°C and 90 °C. The transformation of ammonium dawsonite into Li/Al DHL is not complete at 30 °C and the solid exhibits the absorption bands of both Li/Al LDH and ammonium dawsonite (1543 cm^{-1} and 1451 cm^{-1}). The FTIR spectra of the samples treated at 70 °C and 80 °C are very similar to that of the solid obtained at 60 °C, hence this temperature will be used for preparation of Li/Al LDH.

Li/Al DLH characterization

The FTIR spectrum of Li/Al DLH obtained under the established experimental conditions: 0.5 g ammonium dawsonite, 6.6 mmol Li$_2$CO$_3$, 60 °C and 4 h reaction time, is shown in Figure 5. The FTIR trace shows the characteristic sharp peaks of Li/Al LDH as explained previously.

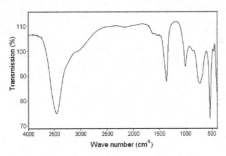

Figure 5. FTIR spectrum of Li/Al LDH obtained at 60 °C, after 4 h reaction time.

33

The XRD spectrum of the sample is presented in Figure 6. The solid was shown to be crystalline. All the reflections observed by XRD could be indexed to aluminum lithium hydroxide carbonate hydrate $Al_4Li_2(OH)_{12}CO_3 \cdot 3H_2O$ (JCPDS 37-185). This result confirms the crystalline nature of the Li/Al LDH sample as determined by FTIR.

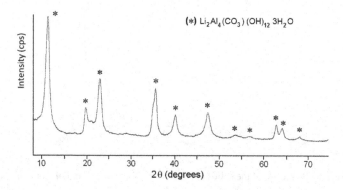

Figure 6. X-ray diffraction pattern of Li/Al LDH obtained at 60 °C, 4 h.

The thermal analysis of Li/Al LDH is shown in Figure 7. The thermal gravimetric (TG) curve evidences five decomposition steps in the temperature range of 25-700 °C. In the first step ranging from 20 to 100°C, the sample releases water absorbed in pores, on the surfaces, etc. This endothermic effect is reflected as a peak on the DTA curvewith a maximum at 96 °C. The elimination of water and carbon dioxide takes place in the range of 100-700 °C and is associated with an endothermic effect at 178 °C in DTA curve. The total mass loss is 43.24%, which is close to the theoretical value of 46.8%.

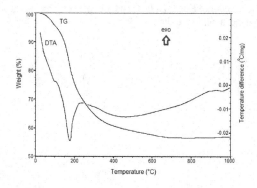

Figure 7. Thermal analysis traces (TGA and DTA) of Li/Al LDH obtained at 60 °C, 4 h.

The bright field TEM micrographs corresponding to ammonium dawsonite and Li/Al LDH are shown in Figure 8. As can be observed, the solid is formed by agglomerates of hexagonal thinly flaky particles with an average diameter of 268 nm. From the particle morphology of both ammonium dawsonite and Li/Al LDH, it can be inferred that the crystallization of Li/Al LDHoccur by a solution crystallization process. The crystallization of Li/Al LDH takes place at pH 10.2. According to solubility curve of aluminum ion, its solubility increase above pH 9.

On the other hand, the Li/Al LDH particles exhibits a more defined morphology than that obtained from basic aluminum sulfate. This fact could be attributed to the absence of impurities present in solution such as sulfate ion.

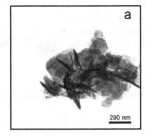 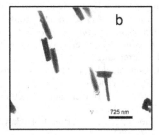

Figure 8. TEM micrographs of: (a) Li/Al LDH obtained at 60 °C and 4 h; and (b) ammonium dawsonite.

CONCLUSIONS

Al-Li layered Double hydroxide has been successfully synthesized using ammonium dawsonite as precursor. The process included the treatment or 0.5 g of ammonium dawsonite with an aqueous solution containing 6.6×10^{-3} mol of lithium carbonate, at 60 °C for four hours. The chemical composition of the synthesized Li-Al LDH corresponds to $[Al_2Li\ (OH)_6]_2\ CO_3$. $3H_2O$, as determined by FTIR, XRD and TA. The TEM image indicates hexagonal flaky particles with an average particle size of 268 nm (diameter). The formation of Li/Al LDH occurs via a solution crystallization process. This preparation method is simple, effective, and the use of ammonium dawsonite as a precursor allows the recycling of lithium carbonate solution without introducing impurities in the process such as sulfate ion.

ACKNOWLEDGMENTS

The authors are grateful to *Yolanda Gallaga O. and Fernando de Jesus Amézquita L.* for their contribution to the experimental work.

REFERENCES

1. L. Ingram, HFW. Taylor, Mineral Mag. **36**, 465-479 (1967).
2. A.V. Besserguenev, A. M. Fogg, R. J. Francis, S. J. Price, D. O'Hare, V. P. Isupov and B. P. Tolochko, Chem Mater. **9**, 241 (1997).
3. M. J. Climent, A. Corma, P. De Frutos, S. Iborra, M. Noy, A. Velty, P. Concepcion, Journal of Catalysis, **269** (1), 140-149 (2010).
4. S. Miyata, M. Taketomi, T. Kojima, S. Tanizaki, A. Hashimoto, K. Kamishiro, M. Eguchi, S. Suzuki, Jpn. Patent No. 2000233188 (29 August 2000).
5. M. Del Arco, E. Cebadera, S. Gutierrez, C. Martin, MJ. Montero, V. Rives, J. Rocha, MA. Sevilla, J PharmSci. **93**, 1649-1658 (2004).
6. F. Li and X. Duan, "Layered Double Hydroxides", ed. X. Duan and D. G. Evans (Springer-Verlag Berlin Heidelberg, 90-109 (2006).
7. C. A. Contreras, A. Padilla, E. Ramos and V. Reyes in *Preparation and Characterization of Aluminum Lithium Hydroxide Carbonate Hydrate obtained from Basic Aluminum Sulfate*, edited by R. Pérez, A. Contreras and R.A. Esparza, (Mater. Res. Soc. Symp. Proc. **1372**, Warrendale, PA, 2012) p. 7-12.
8. C. A. Contreras, S. Sugita, L. M. Torres and J. Serrato, Adv. in Tech. of Mat. Proc. J. (ATM) 36-39 (2003).
9. R. L. Frost, M. L. Weier, and J. T. Kloprogge, Journal of Raman Spectroscopy, **34** (10), 760-768 (2003).
10. J. T. Kloprogge, D. Wharton, L. Hickey and R. L. Frost, American Mineralogist, **87** (5-6), 623-629 (2002).
11. C. J. Serna, J. L. Rendon and J. E. Iglesias, *Clays and Clay Minerals*, **30** (3), 180-184 (1982).

Mater. Res. Soc. Symp. Proc. Vol. 1481 © 2012 Materials Research Society
DOI: 10.1557/opl.2012.1630

Study of Optical Vibrations Modes of Mineral Graphite by Micro Raman Spectroscopy

R. A. Silva-Molina[1], R. Gámez-Corrales[2*], J. M. Hernández-Cazares[3] and I. G. Espinoza-Maldonado[3]

[1] Universidad Autónoma de San Luis Potosí, Doctorado Institucional de Ciencia e Ingeniería en Materiales, DICIM 78290, San Luis Potosí, S.L.P., México
[2] Departamento de Física, Universidad de Sonora, Rosales y Blvd. Luis Encinas 78000, Hermosillo, Sonora, México.
[3] Departamento de Geología, Universidad de Sonora, Rosales y Blvd Luis Encinas 78000, Hermosillo, Sonora, México.
*E-mail: rogelio@correo.fisica.uson.mx

ABSTRACT

We present experimental and theoretical Raman spectra of natural graphite mineral of Sonora, Mexico. In this work, we take the advantage of the utility of the RAMAN spectroscopy as a technique to determine the crystallinity and structure of graphite mineral. The RAMAN spectroscopy provides information that can be used to determine the degree of graphitization, which in turn can be used to know the metamorphic degree of the host rock. The resulting RAMAN spectra of graphite were divided in first and second order regions, in the first region (1100-$1800 cm^{-1}$) the E_{2g} vibration mode with D_{6h} crystal symmetry occurs at $1580 cm^{-1}$ (G band) that indicates poorly organized graphite, additional bands appears in the first order region at $1350 cm^{-1}$ (D band) called the defect band, and another at $1620 cm^{-1}$ (D* band). The second-order region (2200-$3400 cm^{-1}$) shows several bands at ~ 2400 ~ 2700 ~ 2900 $\sim 3300 cm^{-1}$, all of them attributed to electron-phonon interactions or combination scattering. The density functional theory calculations were applied to determine the vibrational properties and the stacking layers of graphite.

Keywords: Microstructure, Amorphous, Geologic, Raman Spectroscopy, electron-phonon interactions.

INTRODUCTION

The carbon atom has electronic configuration $1s^2 2s^2 2p^2$. In bigger structures, the 2s and 2p orbitals are disturbed by nearby atoms and they can form hybrids orbitals sp^2 and sp^3, the resulting orbitals allow up to four covalent bonds. The hybridized sp^2 orbitals form three planar bonding with a 120° angle of separation between them. The atom of carbon linked to other carbon atoms in this configuration results in a hexagonal crystalline network infinitely wide and deep, with a dimensional thickness of an atom. Such a structure is called graphene; the simplest basic component of a graphite crystal is defined as a hexagonal graphene. Graphite is a 3D layering material formed by stacking single layers of sp^2 carbon hexagonal networks. The carbon atoms show strong covalent bonding in a plane and weak van der Waals interactions between planes. The newly discoveries of new classes of nanomaterials such as fullerenes, carbon nanotubes and graphenes, have risen the scientific interest to the study graphite.

In Mexico, Sonora is the third state, after Coahuila and Oaxaca for the occurrence of coal and graphite deposits [1]. Although the graphite production in Sonora date back since 19th century, very little is known about crystallinity and the degree of graphitization, as well as the distribution of these parameters in the graphite belts. Most of the bibliographical references indicate that the graphite in Sonora is of amorphous nature, or very low crystallinity and low temperature of formation (between 200 and 350 °C).

For time, there has been recognized the potential of Raman spectroscopy as a nondestructive technique that provide information about the type of bonding, crystalline structure, atomic valence and conductive properties in minerals or rocks and even in materials with low crystallinity or amorphous [2], which is difficult and in some cases impossibly to obtain by means of other techniques. The Raman spectroscopy technique, widely used by its high sensitivity also has been used as a precise way of quantify the exact number of stacking layers of graphene, size of microcrystal, and defect in samples by the G, D, D* and 2D bands.

In this work, the Raman spectroscopy will be used in determination of crystallinity and graphitization degree of natural graphite from San Jose of Moradillas and San Marcial, Sonora, Mexico. The main vibrational electronic band will be interpreted in terms of the modern Double Raman Resonance theory described by Dresselhaus et al, and Ferrari et al, [3, 4].

The degree of graphitization will be used to determine the maximum temperature reached in the transformation of the carbonaceous strata of Santa Clara Formation in graphite mantos. The maximum temperature of graphitization will help to define the degree of metamorphism reached during the formation of graphite. Finally, the previous results will help to define nature of the solid state of graphite (amorphous or crystalline) in graphite deposits of this part of Sonora, Mexico. Density functional theory helps us to determine the origin of the vibrations obtained of the Raman measurements.

Models and Theoretical Methods

All the calculations were carried out with quantum espresso packages in the preset work using effective core potentials *Ab Initio* density functional theory calculations, DFT,

$E_t = T[\rho] + U[\rho] + E_{xc}[\rho]$, with a plane-wave basis set , spin unrestricted in the general gradient

approximation GGA $\varepsilon xc[\rho] \cong \int \rho(r)\varepsilon xc[\rho(r)]dr$, with Perdew-Wang (PW91) functional was used to obtain all results, geometries optimized and vibrational frequencies of Raman spectra to confirm experimental data.

EXPERIMENTAL PROCEDURE

Samples of graphite flakes were separated by hand from minerals collected from the interior of mines and outcrops. Figure 1 shows some images of the samples collected in the different places. No treatment was done on the samples previously. The analyses were carried out using a microRaman spectrometer X'plora equipment model BX41TF OLYMPUS HORIBA Jovin IVON with a class 3B argon laser with 20-25mW, 532 nm. The operating conditions were at a resolutions of spectral field resolution of 2400gr/mm, in a wave vector range of 150-4000cm^{-1} with dispersion of a slot 6.5cm^{-1}/mm and 50μm. Acquisition time for spectra collection in extended mode, using continuous grating mode. A total of 20 flakes were analyzed for each of the 7 samples, collecting an average of 5 spectra for most of the flake, making a total of 700

Raman spectra. The fitting of the spectrum was performed using Lab Spec software, and Origin 8.0 as a manner of comparison. While the morphology of the flakes were analyzed using a 100x/0.90 (N.A.) optical objective of the inverted microscope.

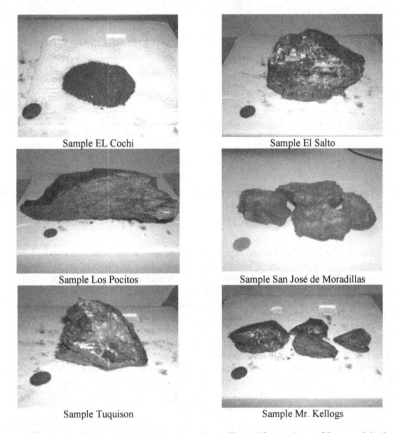

Sample EL Cochi	Sample El Salto
Sample Los Pocitos	Sample San José de Moradillas
Sample Tuquison	Sample Mr. Kellogs

Figure 1. Photography of representative samples collected from mines of Sonora, Mexico.

RESULTS AND DISCUSSION

Figure 2 describes the typical microRaman spectra of natural graphite mineral from Sonora, Mexico. Raman spectrum of graphite is divided in first and second order regions. In the first order region (1100-1800 cm^{-1}) a G band at 1580cm^{-1} indicative of the E_{2g} vibration mode of graphite with D_{6h} crystal symmetry is observed. The so called disorder-induced D band at a 1350cm^{-1} appears frequently in almost all the samples. The D band is attributed to the in-plane

A_{1g} is observed, and is indicative of a disordered and defects in the sample. D* at 1620 cm^{-1} is another Raman band which indices the degree of disorder. Theoretical calculi of DFT allow us to interpret the different vibrations as are showed in figure 2.

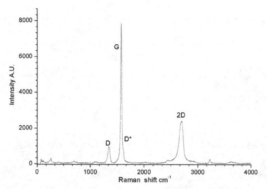

Figure 2. Typical room-temperature Raman spectrum of mineral graphite collected of the region central east of the Sonora state, localized in the Northwest of Mexico, excited by a laser of 532 nm.

Figure 3a shows the first order Raman region of graphite from mines localized nearby San Jose de Moradilla in the central part of Sonora. In all the samples was observed the G band at a shift position of 1580 cm^{-1}, which corresponds to the vibrational mode E_{2g} of a crystal with D^4_{6h} symmetry [5]. This band is typical of high grade crystallinity graphite, the higher the crystal ordering of graphite the more intense and sharper the band, as is observed in samples San Jose of Moradillas, Pocitos, El Porvenir and El Cochi; sample San Jose de Moradillas is the one with more intense and sharper band G (Figure 3). In low ordered materials, band G is wider and less intensity, and is absent in amorphous materials. Although is observed a pronounced reduction in wideness and intensity of Band G in the samples Mr. Kellogs and Tuquison still is the most conspicuous characteristic in the Raman spectra (Figure 3). Nevertheless, the sample El Salto shows the wider and least intensity of all samples, to the grade that this band disappears as a characteristic form the Raman spectra.

Bands D assigned to "crystal defects" in graphite structure are directly related to the magnitude of structural disorder [6-10], these bands are observed in all spectra. Band D appears in position of 1350 cm^{-1} in all the spectra. During graphitization, the relative surface of band D diminishes with the ordering of aromatic layers. The sample San Jose of Moradillas exhibits the smallest band D of all the samples (Figure 3), followed by Los Pocitos, El Cochi and El Porvenir (Figure 3). Band D is more intense and wider in low ordering carbonaceous material, as in samples Mr. Kellogs, Tuquison and El Salto (Figure 3). It is appropriate to mention that these samples show in Raman spectra a big variation in the intensity of band D, and behaves as a weak characteristic, or going so far as to have an intensity higher than band G in the above mentioned spectra. In sample El Salto band D proves to be very wide in most of the spectra with intensity bigger than that of band G (Figure 3).

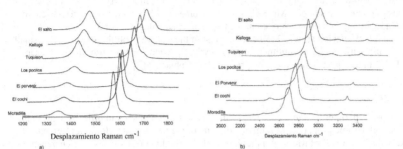

a)   b)

Figure 3. a) First order (1200-1800cm^{-1}) and b) second order (2000-3500cm^{-1}) region Raman spectra of natural graphite mineral graphite from Sonora mines, from bottom to top: San Jose de Moradilla, El Cochi, El Porvenir, Los Pocitos, Tuquison, Kellogs, and El Salto, respectively.

Band D* is located around 1620 cm^{-1} as a small shoulder of band G. The higher the graphitization degree, the lower band D* and decrease up to disappear in highly ordered material. Sample San Jose of Moradillas shows the lowest band D* of all samples, just slightly visible to the right of the band G, like a "little shoulder" (Figure 3). The samples Los Pocitos, El Cochi and El Porvenir shows band D* to the right of the band G like a well-defined small shoulder of very low intensity (Figure 3). In low ordered materials the band D* increases in intensity until it is impossible to separate it from the band G and a wide band forms around 1600 cm^{-1}. The samples Mr. Kellogs, Tuquison and El Salto present a band D* with a third compared to the band G, nevertheless the combination between the bands does occur in any of the spectra (Figure 3). In some analysis of sample El Salto, the resulting spectra exhibits bands D2 with half of the intensity of band G. The band D3 (no showed in Figure 3) is located between the bands D and G, around 1500 cm^{-1}. This band is attributed to defects out of the aromatic plane, such as tetrahedral structures of carbons, organic molecules, fragments or functional groups, commonly in the soot and is related to natural or artificial damages in the structure of the graphite. It is observed as an extremely weak characteristic in the samples Los Pocitos, San Jose of Moradillas, El Cochi and El Porvenir while in the samples Mr. Kellogs, Tuquison and El Salto, the band D3 is almost absent (Figure 3).

Figure 3b shows the second order Raman of the natural mineral. The band 2D is observed around 2735 cm^{-1} in all the samples. As mentioned before, the presence of a "shoulder" of less magnitude corresponds to highly crystalline carbonaceous material, which has reached the order or three periodical structures, which according to Beyssac et al. [6,7] only it appears in graphite that has been exposed to temperatures over 500°C. This feature of band 2D appears most of the spectra of the sample San Jose of Moradillas, while in the samples Los Pocitos, El Cochi and El Porvenir, the band 2D presents just a low incline to the right, but not a "shoulder" (Figure 3b). The samples Mr. Kellogs and El Salto do not show this drop of the band 2D, instead of this some changes in intensity are observed (Figure 3). Band S1 shows big variations in shape and intensity from one spectra o other, so much that in two samples band S1 is observed as a well-developed shoulder.

A new band appears around position of 2900 cm^{-1}, in all the samples as a band of big width and low intensity and interpreted as a product of the combination between first order bands G and D [11], as a consequence of vibrations of the C-H bonds [7, 12] in low crystallinity materials or amorphous graphite. The samples Mr. Kellogs, Tuquison and El Salto are those who present this band with a relatively higher intensity compared with other samples (Figure 3). The band T+D* appears around 2450 cm^{-1} in most of the spectra in all the samples, as a small band of low intensity and high amplitude compared with the band T+D (Figure 3). The presence of band T+D* in the spectra indicates low grade metamorphisms [13].

CONCLUSIONS

In this work experimental and theoretical studies have been conducted in order to analyze the structure and electronic properties of mineral Graphite of the San Jose de Moradilla Region localized in the Northwest of Mexico in the Sonora state. The First order shows the characteristic G (1580 cm^{-1}), D (1340cm^{-1}) and second order Raman region exhibits the 2D (2700 cm^{-1}) bands corresponding to a strongly structure natural graphite. New Raman bands are observed as the results of Raman spectroscopy in mineral natural graphite. These bands only have been observed in graphitic samples produced in laboratory using diverse experimental technique of synthesis. Extraordinaries results give (bands T+D, 2D, T+ D*) the opportunity to produce graphenes or/and carbon nanotubes using natural mineral graphite, opening a new field in the engineering of nanomaterials. Theoretical calculations supported by density functional allowed the interpretation of the experimental results showed above.

ACKNOWLEDGMENTS

This work was supported by CONACyT FOMIX under grant SON-2008-C01-89207 and Rubio Pharma S.A. de C.V. A scholarship from the CONACyT to R.A.S.M. is gratefully acknowledged.

REFERENCES

1. T. J. Corona-Esquivel, R. Benavides-Muñoz , M.E. Piedad-Sánchez, N. Ferrusquía - Villafranca, *Bol.Soc.Geol.Mex.* **LVIII** , 141-160 (2006).
2. P. Sobron Grañon, A. Sanz-Arrans, T.A. García de la Santa, and F. Rull Pérez, *Drainage-Macla,* **10**, 140 (2008).
3. M. S. Dresselhaus, G. Dresselhaus, R. Saito and A. Jorio, *Phys. Rep.* **409**, 47-99 (2005).
4. A. C. Ferrari, J. C. Meyer, V. Scardaci, C. Casiraghi, M. Lazzeri, F. Mauri, S. Piscanec, D. Jiang, K. S. Novoselov, S. Roth and A. K. Geim, *Phys. Rev. Lett.* **97**, 187401 (2006).
5. S. Reich and C. Thomsen, *Phil. Trans. R. Soc. Lond. A*, **362**, 2271-2288 (2004).
6. O. Beyssac, F. Brunet, J. P. Petitet, B. Goffé and J. N. Rouzaud, *Eur. J. Min.* **15**, 937-951 (2003).
7. O. Beyssac, J. N. Rouzaud, B. Goffé, F. Brunet and C. Chopin, *Contr. Min. Petr.* **143** , 19-31 (2002).
8. L. Nasdala, D.C. Smith, R. Kaindl, M.A. Ziemann, Raman Spectroscopy: Analytical Perspectives in Mineralogical Research. A. Beran, A. and E. Libowitzky, (Eds., Emu Notes in Mineralogy: Spectroscopic Methods in Mineralogy. European Mineralogical Union, 2004) pp. 349-422.

9. J. Pasteris and B. Wopenka, *Can. Miner.*, **29**, 1 (1991).
10. F. Tuinstra, J. L. Koenig, *J. Chem. Phys.*, **53**, 1126 (1970).
11. Y. Lee, *J. Nucl. Mat.*, **325**, 174 (2004).
12. R. Tsu, H. González, I. Hernández, *Solid State Commun.*, **27**, 507-510. (1987).
13. T. F. Yui, E. Huang and J. Xu, *J Metamorphic Geol.* **14**, 115 (1996).

Mater. Res. Soc. Symp. Proc. Vol. 1481 © 2012 Materials Research Society
DOI: 10.1557/opl.2012.1631

Chemical Precipitation Synthesis of Nano-Crystalline Mg(OH)₂

A. Medina[1,4,*], L. Béjar[2], G. Herrera-Pérez[3]

[1]UMSNH, Instituto de Investigaciones Metalúrgicas, Edificio U Ciudad Universitaria, C.P. 58040, Morelia, Michoacán, México
[2]UMSNH, Facultad de Ingeniería Mecánica, Edificio W Ciudad Universitaria, C.P. 58040, Morelia, Michoacán, México
[3]Departamento de Ingeniería en Materiales, Instituto Tecnológico Superior de Irapuato (ITESI) Carretera Irapuato-Silao Km. 12.5, El Copal, Irapuato, Guanajuato. C.P. 36821, México
[4]SEP-DGEST-IT de Tlalnepantla, Av. Tecnológico s/n, Col. la Comunidad, Tlalnepantla de Baz, Edo México, 54070, México.
*E-mail: ariosto@umich.mx

ABSTRACT

Magnesium hydroxide (Mg(OH)₂) nanoparticles were synthesized by chemical precipitation synthesis method. The influence of the nano-sized Mg(OH)₂ on the structural modification was evaluated. The formation of Mg(OH)₂ crystals were evaluated by Fourier transform infrared spectroscopy (FTIR) and thermogravimetric analysis (TGA). The particle size and morphology of Mg(OH)₂ was confirmed by high resolution transmission electron microscopy (HRTEM). The crystalline structure of nanoparticles was characterized by fast Fourier transform (FFT) and X-Ray diffraction (XRD), like analytical tools.

Keywords: crystal growth, infrared (IR) spectroscopy, transmission electron microscopy (TEM), Self-assembly morphology.

INTRODUCTION

In recent years, much attention has been focused on the synthesis, characterization and applications in various science and engineering fields of materials with 1D nanostructure.The special attributes of nanomaterials arise due to their unique physical properties like electrical conductivity, optical band gap, magnetic properties and superior mechanical properties, because it exhibits many micro-electronics, optical, electronic, catalytic, nonlinear optical switching and magnetic properties which are different from both their ion and bulk materials [1-3]. The crucial point in obtaining nanosized materials is to prevent particles from aggregating [4,5]. Among various nanomaterials, magnesium hydroxide nanostructure has received a special interest for their attractive scientific and technological aspects in different fields. Magnesium hydroxide is used as halogen-free flame retardants for polymers, electric cables, building and decoration materials [6], preparing foodstuff starch esters [7] and treating wastewater [8]. With the emergence and the increase of microbial organisms, many researchers have been focusing on the new and effective antimicrobial reagent developments. Inorganic antibacterial agents received more and more recognition as antibacterial products [9,10]. Inorganic nanoparticles, whose structures exhibit novel physical, chemical and biological properties, have attracted much interest. Among the inorganic antibacterial agents, $Mg(OH)_2$ is well known as an effective antibacterial agent [11]. Currently, the synthesis of $Mg(OH)_2$ nanomaterials is being carried out by different techniques to achieve desired size and shape. Different methods, such as solvo-thermal method [12], reverse precipitation [13], hydrothermal route [14] and solution based chemical process [15] were adopted by scientists to prepare $Mg(OH)_2$ nanomaterials. Thus, the synthesis of inorganic nanocrystals with controlled size and shapes are of special interest in nanoscience and nanotechnology, because it is now widely accepted that many fundamental properties and applications of nanocrystals depend strongly on their shape, size and size distribution [9,10]. In this paper we describe a simple method to synthesize $Mg(OH)_2$ nanoparticles using the chemical precipitation method from inexpensive starting materials like sodium hydroxide. The properties of product were investigated and the formation mechanism was discussed on the basis of our experimental results. In this communication, we report the characterization of high quality arrays of $Mg(OH)_2$ crystals synthesized by a reproducible synthesis. Three cases were considered, the limiting reagent (1:1), the reaction stoichiometry (1:2) and the reagent excess (1:3).

EXPERIMENTAL PROCEDURE

In the reaction a-$MgSO_4 \bullet 7H_2O_{(aq)} + b$-$NaOH_{(aq)} \rightarrow Mg(OH)_{2(s)} \downarrow + 2Na\ SO_{4(aq)}$, a and b are the stoichiometric coefficients, the relation for each case considered in this work will be indicated as $a{:}b$. In the first step 40 g of $MgSO_4 \bullet 7H_2O$ was dissolved in 200 mL of distilled water, and stirred at 475 rpm a constant temperature of 40 °C. A second solution was prepared by dissolving 7.3 g of NaOH in 200 mL of distilled water and added to the Mg-solution with a speed of 4 mL/min and stirred for 30 min. The precipitates were centrifuged and washed in order to eliminate the Na^+ and $SO_4^=$ species, and dried at room temperature for 20 h. Different analytical tools were used to characterize the metal hydroxides synthesized. FTIR spectra were collected on a Perkin-Elmer spectrophotometer model Spectrum RX-I by transmitting technique and dispersion in KBr pellet formation by compression. TGA analyzes were performed on an equipment TGA-DTA mark TA-Instrument model STA 409, the heating rate was of 10 °C/min with a dynamic atmosphere (20 mL / min) of air. XRD analysis of the phases and crystallographic structures of the samples was carried out using a Siemens D5000 X-ray diffractometer equipped with graphite monochromatized high-intensity CuK_λ(λ=1.54178 Å). The Bragg angle 2θ ranges from 15° to 75° at a scanning rate of 0.06°/s. Further, the morphology and particle size of $Mg(OH)_2$ was confirmed by recording SEM using a JEOL JSM 6400 microscope and HRTEM using a FEG TECNAI F20. Due the $Mg(OH)_2$ crystals are a semiconductor material it was necessary to cover them with a copper layer before to carry out the SEM analysis. The samples for HRTEM analysis were prepared by spreading a droplet of solution of $Mg(OH)_2$ particles onto a carbon film supported by a Cu grid and subsequent drying in vacuum.

RESULTS AND DISCUSSION

The structure of $Mg(OH)_2$ was confirmed by FTIR spectroscopy. Figure 1 show three important characteristics peaks. The peaks appeared from 300 to 1250 cm^{-1} account for the metal hydroxide stretching vibrations. A Peak at 1629 cm^{-1} explains the bending vibrations of water molecules attached with the Mg(OH)$_2$. The vibration of the O-H bond stretch is around 3700 cm^{-1}, this signal is particularly strong and sharp Mg(OH)$_2$.

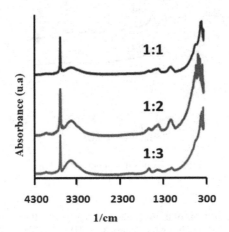

Figure 1. FTIR spectrum of Mg(OH)$_2$ at different molar ratio.

Thermal stability of Mg(OH)$_2$ (Figure 2) was analyzed by recording TGA under air atmosphere at the heating rate of 10 °C/min. The thermogram showed a three-step degradation process. A step at 445°C explains the dehydroxylation process with the removal of water molecules i.e., at elevated temperature, the hydroxyl groups of magnesium hydroxide condensed with each other and remove the water molecules.

Above 600 °C it showed that the % weight of Mg(OH)$_2$ retained was 48%. Due to the amino intercalation into the Mg(OH)$_2$ structures, the thermal stability of char formation above 600 °C was increased. At low temperature (below 300 °C), Mg(OH)$_2$ system shows higher thermal stability. Redform and Fathima Parveen [16,17] synthesized the Mg(OH)$_2$ and they have reported the TGA of Mg(OH)$_2$. Our report is in accordance with Redform and Fathima[16,17].

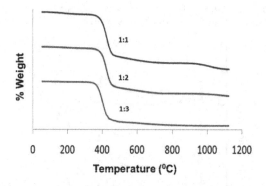

Figure 2. TGA profiles of pure $Mg(OH)_2$.

Figure 3 show the XRD patterns of the hydrothermal prepared $Mg(OH)_2$ samples.All the diffraction peaks can be indexed to hexagonal phase $Mg(OH)_2$(JCPDS # 00-088-0206). It was noticeable that the patterns exhibits a strong intensity of the (011) peak, this results strongly indicate the preferential growth of $Mg(OH)_2$ crystals. In addition, the lattice constants of the as-crystallized array are measured to be a=0.3143 nm and c=0.4751 nm, which are in good agreementwith the values of the brucite ($Mg(OH)_2$) of the space group P-3m1 (164), which is expected to occur in this way given the thermodynamic nature of the phase. The samples with molar ratio of 1:1 and 1:2 are rather broad, indicating the nanocrystalline size. No characteristic peaks from other impurities were detected. The sharpness of the diffraction peaks implies a high crystallinity of the $Mg(OH)_2$ samples.

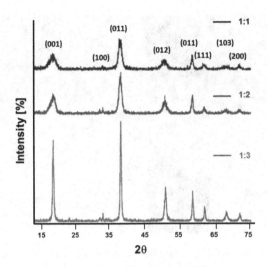

Figure 3. XRD patterns of ZnO crystals synthesized a different molar ratio. (The peak indexes are specified above the peaks).

HRTEM images of $Mg(OH)_2$ are showed in Figures 4, 5 and 6 for molar ratio of 1:1, 1:2 and 1:3 respectively. All figures show different crystals planes, morphology and size. Figure 4a shows the formation of spherical shape nanoparticles in the range from 10 to 5 nm of size, in figure 4b we can appreciate the FFT image which represents a policrystal image in the (011) plane which correspond an interplanar distance of .0236 nm. Figure 5a infers the nanorod-like morphology. This nanorod can be better observed in the figure 5b, but it gets hydrolytically oxided and forms an amorphous structure. The formation of small nanorod crystals produce a circular polycrystalline image which is showed in figure 5c which correspond at (011) plane whit interplanar distance of 0.236 nm. The width of the nanorods is determined < 6 nm and the length > 40 nm.

Figure 6a infers an idea about the layered crystals planes of $Mg(OH)_2$. Hence the molar ratio 1:3 acts as a self-assembly and produces structurally perfect crystals of $Mg(OH)_2$. Also, it is generally acceptable that preferential absorption of organic molecules in the solution to different crystals faces directs the growth of nanoparticles into various shapes by controlling the growth rates of the crystals faces or directions [18]. Currently, Wu [19] reported the synthesis of superfine $Mg(OH)_2$ with the size of 250 nm in spherical shape. In our system, the size of $Mg(OH)$ is > 10 nm for spherical shape morphology and the width < 6 nm and length < 40 nm for nanorod-like morphology. Our results coincide with their reports.

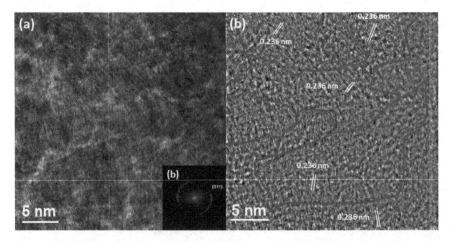

Figure 4. a) HRTEM image of the Mg(OH)₂nanoparticles synthesized with molar ratio of 1:1 b) FFT poly-crystal image showing the (011) plane from image 4a.

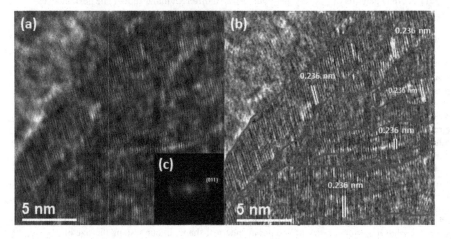

Figure 5. a) HRTEM image of the Mg(OH)₂ nanorods-like morphology synthesized with molar ratio of 1:2, b) IFFT image showing the nanorods and c) FFT image of the nanorods in the (011) plane

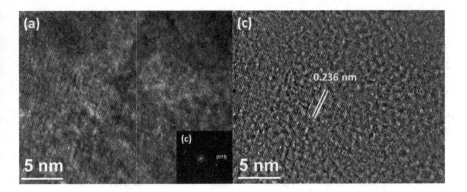

Figure 6. a) HRTEM image of the Mg(OH)$_2$ nanoparticles synthesized with molar ratio of 1:3 with a interplanar distance of 0.236 nm, b) FFT image showing the (011) plane from image 6a.

CONCLUSIONS

In present work, Mg(OH)$_2$ was synthesized with different molar ratios and the time by the crystal growth was also performed by chemical precipitation method and have been characterized by FTIR, TGA, XRD and HRTEM. A facile route for the synthesis of Mg(OH)$_2$ crystals has been produced and reported. This research obtains evidence of effect of residence time controlled by the precipitating reagent addition. Monodisperse single crystalline Mg(OH)$_2$ self-assembly (1:3 molar ratio), spherical particles (1:2 molar ratio) and nanorods-like morphology (1:2 molar ratio) were produced in large quantities. Nanosized metal hydroxide analysis was confirmed by HRTEM. All crystals produced showed a higher thermal stability. According to FFT analyses, the Mg(OH)$_2$ crystals had brucite orientation. The nuclei and subsequent development on surfaces (011) were observed in the nanorod. The method offers a simple cost path and low for the production of magnesium hydroxide matrices.

REFERENCES

1. S. Nie and S. R. Emory, *Science,* **275,** 1102 (1997).
2. C. Wang, N. Niflyn and R. Larger, *Adv. Mater,* **16,** 1074 (2004).
3. A. N. Shipway and I. Willner, *Chem. Commun.* **20,** 2035 (2001).
4. R. F. Ziolo, E. P. Giannelis, V. Mehrotra and D. R. Huffman, *Science,* **257,** 219 (1992).
5. M. Buemann, H. Colfin, D. Walsh and S. Mann, *Adv. Mater.* **10,** 237 (1998).
6. Y. Ding, G. Zhang, H. Wu, B. Hai, L. Wang and Y. Qian, *Chem. Mater.* **13,** 435 (2001).
7. Anheuser Busch Inc., US Patent US3839320-A, (1974).
8. Mitsubishi Rayon Co. Ltd. (MITR-C). Japan Patent P52105653-A, (1977).
9. X. J. Wang, X. L. Qiao, J. G. Chen, H. S. Wang and S. Y. Ding, *J. Ceram.* **24,** 39 (2003).

10. M. Fang, J. H. Chen, X. L. Xu, P. H. Yang andH. F. Hildebrand, *Int. J. Anti-microb. Agents,* **27,** 513 (2006).
11. C. Dong, D. Song, J. Cairney, O. L. Maddan, G. Heand Y. Deng, *Mat. Res. Bul.* **46,** 576 (2011).
12. Y. Li, M. Sui, Y. Ding, G. Zhang, J. Zhuang and C. Wang, *Adv. Mater.* **12,** 818 (2000).
13. J. C. Yu, A. W. Xu, L. Z. Zhang, R. Q. Song and L.Wu, *J. Phys. Chem.* **B108,** 64 (2004).
14. C. L. Yan, D. F. Xue, L. J. Zou, X. X. Yan and W. Wang, *J. Crys. Grow.* 282, 448 (2005).
15. S. M. Zhang and H. C. Zeng *Chem. Mater.* **21,** 871 (2009).
16. S. A. Redform and B. I. Wood, *Am. Miner.* **77,** 1129 (1992).
17. M. Fathima Parveen, S. Umapathy, V. Dhanalakshmi, R. Anbarasan. *Nano Brief Reports and Reviews,* **4,** 147 (2009).
18. C. J. Murphy, *Science,* **298,** 2139 (2002).
19. X. F. Wu, G. S. Hu, B. B. Wang and Y. F. Yang, *J. Cryst. Growth,* **310,** 457 (2008).

Mechanical and Microstructural Characterization of Steels used in the Oil Industry

Mater. Res. Soc. Symp. Proc. Vol. 1481 © 2012 Materials Research Society
DOI: 10.1557/opl.2012.1632

Effect of the Aging Treatment in Micro-Alloyed Steel

M. A. Doñu Ruiz[1*], J. A. Ortega Herrara[1], N. López Perrusquia[1], V. J. Cortés Suárez[2],
L. D. Rosado Cruz[3]

[1]Instituto Politécnico Nacional, Grupo e Ingenieria Mecánica Computacional, SEPI-ESIME,
Adolfo López Mateos, Zacatenco, México D.F., 07738, México.
[2]Universidad Autónoma Metropolitana UAM unidad Azcapotzalco, Departamento de Materiales,
Av. San Pablo 180 Azcapotzalco, Mexico .DF, 02200;
[3]Universidad Politécnica del Valle de México UPVM. Av. Mexiquense s/n esquina Av
universidad politécnica, Col villa esmeralda, 54910, Tultitlan. Edo de México.
*E-mail: mdonur0800@ipn.mx

ABSTRACT

This work study the effect on aging thermal treatment on micro-alloyed steels API X70 pipe,
microstructure and mechanical properties such a yield strength *(Y)*, hardness *(Hv)* and Young's
modulus *(E)* are presented in this work. Thermal treatment consists of two phases: i) The
solution treatment introducing samples in a electric induction furnace at 1100 °C for 30 min
under argon atmosphere and water quenching, ii) aging process for five temperature in the range
between 204 to 650 °C for 30 min of time exposition and water quenching, respectively. The
microstructural characterization was examined by optical microscopy and matrix samples aging
showed microstructures like acicular ferritic, polygonal ferritic and bainitic-ferritic, and the
secondary phases were examined by scanning electron microscopy (SEM) and energy dispersive
spectroscopy (EDS) obtained by SEM evidencing the presence of precipitates composed of
vanadium *(V)*, niobium *(Nb)* and titanium *(Ti)*. The mechanical properties were evaluated by
depth sensitive indentation test at the samples aging, the results showed increase of the *(Hv)* and
(E) to the conditions of low temperature aging.

Keywords: microstructural, hardness, second phases, nano-indentation, scanning electron
microscopy (SEM).

INTRODUCTION

High strength low alloy (HSLA) steels or microalloyed steels are a technologically important
structural materials in application included pipelines (of crude oil or natural gas transport), body
parts in automotive industry and pressure vessels [1]. A lot of researches have dealt with the
acicular ferrite dominated microstructure and its properties [2,3].

These steels contained small amounts of alloying elements, such as titanium, niobium and
vanadium, which enhance the strength through the formation of stable carbides, nitrided or
carbonitrides are important in the control of austenite recrystallization.

The mechanical characterization of the materials is represented by hardness and is defined as
the resistance of permanent deformation. In depth sensing indentation, hardness H is defined as
the mean contact pressure P_m under the loaded indenter, $H=P/A= P_m$ and is calculate as the
indenter load P divided by the projects contact area A. Yield strength Y can, be calculate from the
mean contact pressure P_m (or hardness). For soft metal with low hardness compared to elastic
modulus, yield stress is usually calculated according to Tabor [4] as $H=CY$, where the H is the

hardness, C is a constant and Y is the representative stress, concluded that C is from 2.8 to 3 for coppers and mild steels. Jhonson`s study [5] proposed the relationship between hardness and yield strength.

$$p_m / Y = A(B + In(k(E / Y)\varepsilon_{rep})$$ (1)

Where A, B and k are constants (A=4/3, B=2/3 and k=5/3). Actually, the right-hand part of Eq. 1 corresponds to the constraint factor C, which depends on ε_{rep} and the ration E/Y that characterises the material susceptibility to plastic flow.

The values of representative strain ε_{rep} for Vickers and Berkovich indenter vary from 0.0115 to 0.042. In this study, the behavior of the material beneath the pyramidal indenter was assumed to be a plane strain deformation and ε_{rep} is defined as value of 0.033, values obtained by Ogasawara [6].

In the present work, the mechanical properties of API X70 microalloyed steel subjected to aging treatment, and the relation between microstructure and evaluation of the mechanical properties by depth sensitive indentation test was discussed.

EXPERIMENTAL

Microalloyed steel plates 37.5x2.5x7mm which belong to API X70 steel according to API 5L specification were used as specimen with composed of 0.027% *C*, 1.51% *Mn*, 0.13% *Si*, 0.014% *P*, 0.002% *S*, 0.035% *Al*, 0.093% *Nb*, 0.28%*Cu*, 0.27% *Cr*, 0.16% *Ni*, 0.004% *Mo*, 0.004% *V*, 0.011% *Ti*, 0.16% *N* and *Fe* balance and in the Figure 1a show the microstructure in direccion hot rolling. Thermal treatment consists of two phases: i) The solution treatment introducing samples in a electric induction furnace at 1100 °C for 30 min under argon atmosphere and water quenching, ii) Aging process at temperature 204, 315, 426, 538, and 650 °C for 30 minute and water quenching, respectively , according to the diagram presented in Figure 1b. The specimens were cross-sectioned along the rolling direction, mechanically ground and polished by conventional metallographic techniques and etching with 2 to 5% nital solution. ZEISS AX10 optical microscopy and JSM-6010LA scanning electron microscope (SEM) were used to examine microstructure characterization. Also, the EDS evidenced the presence of precipitates.

Depth sensitive indentation test were carried out using a dynamic ultra microhardness tester DUH-211S, the sample were indented at room temperature with a Berkovich diamond with 115°C triangular pyramid indenter, the peak load applied is 100 mN to estimate the elastic modulus and microhardness of the constituent phases.

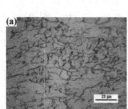 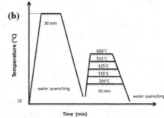

Figure 1. a) View of the microstructure parallel to the rolling direction in API X70 steels, 100X and b) schematic of diagram of the aging treatment processing.

RESULTS AND DISCUSSION

Microstrcutural characterization

Figure 2 shows the optical and scanning electron microstructure of API X70 after aging.

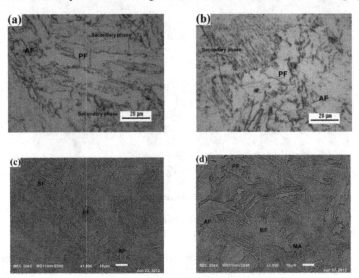

Figure 2. Microstructure obtained by optical microscopy at the temperature aging of a) 204 °C and b) 650 °C and SEM microstructure to temperature aging of c) 204 °C y d) 650 °C

Optical microstructure show in Figure 2a and Figure 2b at the temperature aging of 204°C and 650°C, respectively, reveals mixed morphology of ferrite grains (α) along with the distribution of low dark color phases at the ferrite grain boundaries, the correspondent SEM micrograph shown in Figure 2c and Figure 2d at the condition aging of 204°C and 650°C, respectively reveals island with brighter contrast and show phase with featureless appearance for polygonal ferrite (PF), acicular ferrite (AF), martensite and retained austenite (MA).

Higher magnification obtained by SEM revel second phases on API X70 aging, in figure 3 at the condition 204°C, show the secondary phases of islands degenerate perlite (DP), bainite ferrite (BF). In this result, MA and BF includes many carbide particles.

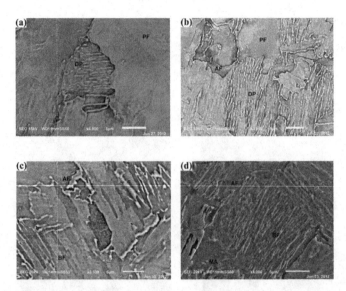

Figure 3. Higher magnifications obtained by SEM, micrographs reveals second phases distributed between ferrite grains at the condition 204°C aging.

The microstructure on API X70 consist acicular ferrite (see Figure 1a), Figure 3 shows the microstructure at the condition 204°C aging. The ferrite grains are smooth, and the equiaxed ferrite grains separated by continuous, linear boundaries. The dark etching areas of the microstructure consist of martensite that formed in untransformed austenite (γ) during quenching. Hence, at the condition 650 °C again, shows a slightly coaserned ferrite microstructure as some ferrite were grown the volume fraction of PF (see Figure 2b and Figure 2d) increase with increase the temperature aging. So, some precipitation strengthening by Nb (C, N), occurs during the $\gamma \rightarrow \alpha$ transformation during cooling.

EDS by scanning electron microscopy

Figure 4 show the plot of counts versus KeV obtained during SEM microanalyses carried out in areas precipitates on API X70 aging treatment at 204°C, the type of precipitate evidence, which vary in size from several microns.

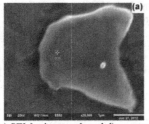 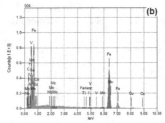

Figure 4. a) SEM micrograph and d) corresponding EDS spectrum on API X70 aging at 204°C.

The microstructure observed the presence of carbides in the matriz ferrite Figure 4a. according the figura 4b, the presence of the element such as Nb, Ti, V, C and N, EDS analyses of particles were performed and the chemical composition is listed in Table I. The presence precipitation of the fine (Nb,Ti)(C,N) from the ferrite generate at low temperature resulted in effective precipitation strengthening effect.

Table I. Chemical composition for the analysed spectra in mass fractions.

Element	C	N	Ti	V	Mn	Fe	Cu	Nb	Mo
ms%	13.54	27.43	0.16	0.15	0.79	57.06	0.36	0.16	0.4

Depth Sensitive indentation

Figure 5 shows the response of the aging treatment to depth sensitive indentation test with different phase with constant load. The indentations were obtained after aging at 538 °C, the elastic modulus could be calculate from the slope of the load displacement curve using the Oliver and Pharr equations [7] and figure 5d show the behavior of the load displacement of the constituent phases.

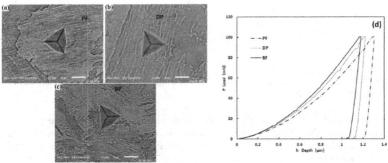

Figure 5. SEM images of nanoindentation of differents phases a) PF, b) DP and c) BF and d) load-displacement curves on different phases.

Figure 6a show the behavior of the load-displacement obtained in the phase PF and the Figure 6b show the results the microhardness obtained to different temperatures aging.

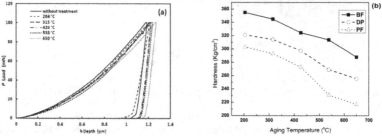

Figure 6. a) Load-displacement curves on API X70 aging at different temperature on the phase PF and b) variation of microhardness on different phases, with different temperature aging conditions.

Figure 7 present the mechanical properties related to the homogeneity associated with the single phase microstructure develop by the aging treatment, Figure 7a and 7b show aging curve plotted modulus elastic and yield strength between aging temperature, respectively. The yield strength was obtained by equation 1, depth sensitive indentation test was carry out on DP, BF and PF with constant load applied, the load-displacement curves with distinctly different slopes during the load cycle.

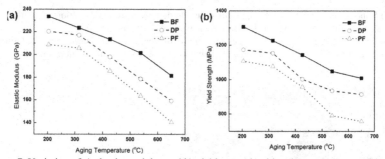

Figure 7. Variation of a) elastic modulus and b) yield strength with aging temperature after water quench on API X70.

The variation of mechanical properties values with regard to aging temperature and constituent phases BF, DP and PF. The elastic modulus of the concerned phase are in range of 181 to 233GPa (BF), 159 to 220 (DP) and 149 to 208 GPa (PF) with an average error margin of 4 GPa. The variation of the microhardness change from 287 to 354 Kg/cm² (BF), 255 to 321 Kg/cm² (DP) and 216 to 302 Kg/cm² (PF) with an average error margin of 15 Kg/cm² and the variation of the yield strength obtained are range 1008 to 1306 MPa (BF), 913 to 1173 MPa (DP) and 757 to 1107 MPa (PF) with an average error margin of 70 MPa. According Thompson *et* al.

[8] the mechanical properties of low-carbon steels with acicular ferrite/MA microstructures showed that yield strengths ranged from 450 to 985 MPa. Hence, the molybdenum (Mo) contained steels exhibited higher strength and the yield strength increased. Nb and Ti additions play an important role in the microstructure formation [9].

CONCLUSIONS

The present study evaluate the effect of the aging treatment in API X70 micro-alloyed steel, the aging significantly affects the properties of the material and depth sensitive indentation is a powerful technique for evaluate the mechanical properties of materials. The mechanical properties related to the heterogeneity are associated with the multiphase microstructure develop by the aging treatment. The heat treatment temperature of 204 °C increase the properties and micro alloyed steels without treatment having a yield strengths, report in literature about 690 MPa, and the present work by aging treatment are obtained values of 757 to 1306 MPa.The EDS evidence the formation of carbonitrures (Nb,Ti) C,N and the change the mechanical properties. The presence of Nb, Ti (C, N), lead to changes in the mechanical behavior of the material: the yield strength and the microhardness increase.

ACKNOWLEDGEMENT

The authors gratefully acknowledge the CONACyT of Mexico City.

REFERENCES

1. J. Davis, *High-Strength Low-Alloy Steels.* ASM International, second edition ed.,1990.
2. M. Milizer, Warren J. Poole. *Material Technology,* Steel Research, **69,** (1998).
3. R. W. K. Honeycomb, London Edward Arnold Publishers (1998).
4. D. Tabor *The Hardness of Metal* published in the Oxford classics series. New York: Oxford University Press; 2000.
5. K. L. Johnson *Contact Mechanics,* UK: Cambridge University Press, (1985).
6. N. Ogasara, *J. Mater Research,* **20,** 2225-2234, (2005).
7. W.C. Oliver, G.M. Pharr, *J. Mater. Research,* **7,** 1564 (1992).
8. S.W. Thompson and G. Krauss, Mechanical Working and Steel Processing; Proceedings of the Iron and Steel Society, 467–481, (1989).
9. T. Abe, K. Tsukada, I. Kozasu, in: J.M. Gray et al., (Ed.), *Proceedings of International Conference on HSLA Steels:* Metallurgy and Applications, Beijing, p. 103, (1985).

Mater. Res. Soc. Symp. Proc. Vol. 1481 © 2012 Materials Research Society
DOI: 10.1557/opl.2012.1633

Characterization Microstructural and Mechanical of X-60 Steel Heat-Treated

N. López Perrusquia[1*], J. A. Ortega Herrera[1], M.A. Doñu Ruiz[1], V. J. Cortes Suarez[2], L. D. Cruz Rosado[3].

[1]Instituto Politécnico Nacional, SEPI-ESIME, Adolfo López Mateos, Zacatenco. México.D.F. 07738. México.
[2]Univerisdad Autónoma Metropolitana Unidad Azcapotzalco, Av. San Pablo 180 Azcapotzalco 02200, México .D.F. Área de Ciencia de los materiales.
[3]Universidad Politécnica del Valle de México, Grupo Ciencia e Ingeniería de Materiales. Av. Mexiquense, Tultitlán. Edo. México. México.
*E-mail: nlopezp0803@ipn.mx

ABSTRACT

In this paper was study the effect of heat treatment on mechanical properties of an API X-60 steel used for storage and transportation of hydrocarbons. In the first stage evaluation are mechanical properties of steel heat treated by the technique of the three-point test according to ASTM 399-90 was carried out. In the second stage, characterization of the type of failure and microstructure through optical microscopy (OM) was determined; also heat treated samples were then mechanically tested for hardness (HRC) and nano-indentation. The presence of alloying elements by scanning electron microscopy (SEM) and the fracture surfaces generated in the steel with varying times, temperatures and cooling medium generated by different modes of solicitation (Bending), likewise with loading rates were determined. The results revealed a ductile fracture and microstructures (PF) ferrite-pearlite (DP), bainite -ferrite (BF) and martensite-retained and martensite/retained austenite (MA). Finally, this article discusses the effect of heat treatment followed by precipitation hardenable of steel API X-60 on the mechanical properties.

Keywords: steel, optical metallography, scanning electron microscopy (SEM), hardness, nano-indentation

INTRODUCTION

Generally, heat treatment improves steels toughness and hardness, and it is absolutely necessary for the proper functioning of steels. The heat treatment usually consists of austenitizing, quenching and is followed by multiple tempering. After this procedure the steels gain properties that are suitable for industrial applications [1-2]. The steels specified API X-60 in the API standard (American Petroleum Institute), are mainly are used in the oil industry, where the principal alloying elements added to steel in widely varying amounts either singly or in complex mixtures are nickel, chromium, manganese, molybdenum, vanadium, niobium, silicon and cobalt; these alloying elements permit Precipitation hardening of steels API X-60 for hydrocarbons and storage are being proposed for such an application, yielding improved mechanical properties and increased life-time. Also the type of microstructure of the steel pipe is a significant feature that is required to increase its mechanical, physical and chemical properties. The test material is commercial steel API X-60, this type of steel is hot rolled, which is subsequently rapidly cooling; a structure obtained as austenite, pearlite and cementita [3-4]. Furthermore steel for transport and storage require better microstructural characteristics and mechanical properties for increased service life therefore this study aims to study the microstructural changes with conventional thermal treatments to observe their behavior in the mechanical properties of this material.

EXPERIMENTAL

Materials

The studied steel API X-60 with chemical composition shown in table I, with microstructure comprising ferrite and perlite, typical of these steels is shown in figure 1.

Table I. Chemical composition of studied material (wt%).

Steel	C %	Mn%	Si%	P %	S %	Cr%	Ni%	V %	Nb%	Ti%
API X- 60	0.21	1.52	0.19	0.012	0.003	0.16	0.15	0.05	0.03	0.01

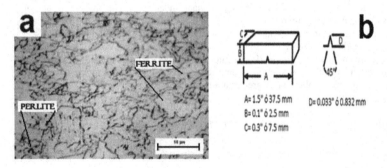

Figure 1. (a) Micrograph of API-X60 steel base and (b) Test specimen of three points.

Heat treatments

The API X60 steel were heat treated at 1000 ° C, with inert argon atmosphere, with 20 minutes to solubilize; retention thereafter of treatments at 204, 315, 426, 538 and 650 ° C for 30 minutes; with water quench.

Metallography

Microstructural characterization and fracture type present was determined by a metallographic microscope Olympus GX 51 etched with 2 % nital and observed by scanning electron microscopy JEOL 6063 L (SEM) to reveal the morphology of the phases.

Hardness and nano-indentation

Heat-treated samples were tested for various mechanical properties. Average HRC readings were determined by taking five hardness readings at different positions on the samples, using a CV-Instrument - 700 universal hardness tester and treatment using a Mitutoyo MXD200 computer also was completed with the instrumented nanohardness Ultra Micro Hardness Tester Mitutoyo

Technique of three-Point

Bending tests were performed on a Universal Mechanical Testing Machine, Instron brand model 8502, with capacity of 20 kN, with feed rate 0.102 mm / min.

RESULTS AND DISCUSSION

The microstructural characterization by optical microscopy is shown in the figure 2 where observed ferrite structure proeutectoide white, also with dark areas perlite also the formation of martensite and bainite a similar result observed by C. Hurtado et. al. and Wei Wang et. al. Furthermore after the steelwork heat treatment, were obtained a microstructure is homogeneous bainite. Small randomly distributed carbide particles, not completely resolved by optical microscopy, are present in the bainitic matrix, probably Mo, Ti, Nb and V carbides.

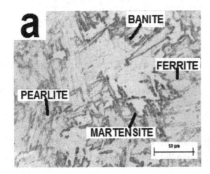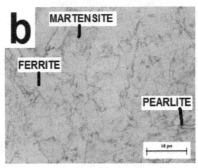

Figure 2. Microstructure of API X-60 consisted of ferrite-pearlite, banite and martensite-colonies (20X): (a) 625 ° C and (b) 204 ° C

Moreover in the micrographs shown in figure 2 is a higher concentration of the ferrite phase. Likewise, the perlite which is found in these steels depends on several variables such as the initial size of retained austenite, before transformation γ / α also the chemical composition of steel and the cooling rate during treatment [7-8]. Scanning electron microscopy shows perlite structure with a grain size of ferrite and retained austenite. Figure 3 shows a magnification of 800X of the ferrite grain shape irregular proeutectoide [9-10]. Furthermore have perlite grains with different orientation and with a very irregular layer of cementite, besides martensite areas. Likewise an EDS were performed where the alloying elements are disclosed which tend to form TiN, NbC, Ti$_2$S, among others, as shown in figure 3.

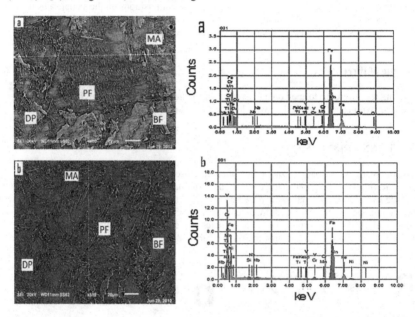

Figure 3. Micrographs of API - X60; etched with 2%; (a) 204 ° C and (b) 625 ° C; polygonal ferrite grains (PF) and in some regions, islands of second phases with degenerate pearlite (DP), bainite ferrite (BF) and martensite/retained austenite (MA) and spectrum of the analysis the area of mapping (a) 204 ° C and (b) 625 ° C.

Figure 4 (a) shows a change of hardness with each treatment, where treatment at 204°C shows a greater increase in comparison to other temperatures in these cases [11-12]. Figure 4 (b) illustrates the nano-indentation tests.

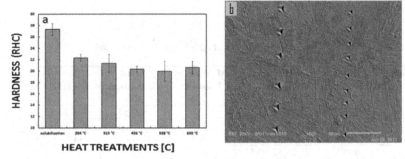

HEAT TREATMENTS [C]

Figure 4. (a) Schematic diagrams showing the data of hardness (HRC) API- X60 steel heat treated, (b) Profile of nanoindentation of API-X60 steel whit specimen 204 ° C.

Figure 5 (a) shows that there an increase in the load of the specimen at 204 ° C, followed by 315 ° C, subsequently have a low hardness of the other specimens heat treated. It also presents ductile fracture type for this steel with the heat treatment applied, also precipitation observed in the fracture zone figure 5 (b).

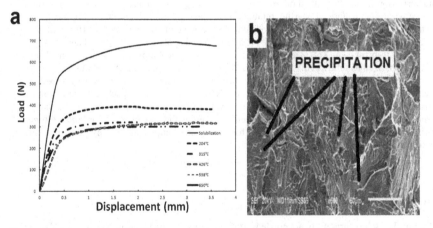

Figure 5. (a) Curves load-displacement of heat-treated specimens (b) type of fracture of the specimen at 204 ° C.

Table II presents the mechanical properties obtained with the treatments proposed in this paper [13]. Likewise shows the effects of heat-treated on the hardness, modulus of elasticity and load maximum.

Table II. Mechanical properties of API-X60 steel heat treated.

Materials	Heat Treatment	Hardness (HRC)	Modulus of Elasticity (Gpa)	Load Máximum (N)	Hardness (Hv)
X-60	solubilization	27.23 ±1.40	255	693.57	340.34 ±2.23
X-60	204°C	22.30 ±0.59	215	321.38	290.23±1.34
X-60	315°C	21.32 ±1.58	207	319.64	244.23 ± 2.13
X-60	426°C	20.32 ±0.52	200	315.78	237.34 ± 1.98
X-60	538°C	20.60 ±1.74	198	302.29	228.78 ±1.12
X-60	650°C	19.95 ± 1.14	193	297.67	223.12 ±1.08

CONCLUSIONS

In this paper we study the influence of the heat-treatment on API-X60 steel. Microstructural characterization studies by optical microscopy and scanning electron microscopy show a ferrite structure with small islands of pearlite, retained austenite also have been found on this steel. This steel also showed a brittle fracture. Likewise there is a change in mechanical properties with the heat treatments described in this paper, The results obtained applying constant load and slow strain rate of tests on API- X60 steels having different mechanical properties of heat-treated and can be summarized as follows: the treatment at 204 ° C presents a change in modulus of elasticity, maximum load and hardness greater than the other treatments. The typical EDS spectrum presents to alloying elements presents in steel API-X60. Moreover they reached the precipitates formed on the surface, thus it is expected that this work give emphasis for future work on the hydrogen embrittlement, because these treatments is to allow the precipitates retard the hydrogen embrittlement of these materials that are used for storage and transportation of hydrocarbons, in addition to storage of bio-fuels.

ACKNOWLEDGEMENTS

Two of the authors gratefully acknowledge the support given by Consejo Nacional de Ciencia y Tecnologia, CONACyT through the Doctoral Program of Instituto Politecnico Nacional and Promep.

REFERENCES

1. Wei Wang et. al. *Mater. Sci. and Engin.* **A 502**, 38–44 (2009).
2. C.Y. Chen, H.W. Yen, F.H. Kao,W.C. Li, C.Y. Huang, J.R. Yang, S.H.Wang. *Mater. Sci. and Engin.* **A 499**, 162–166 (2009).
3. G Ananta Nagu, Amarnath, T K G Namboodhiri, *Bull. Mater. Sci.* **26**, 435–439 (2003).
4. S.H. Salleh, M.Z. Omar, J. Syarif, S. Abdullah, *J. of Sci. Res.* **34**, 83-91 (2009).
5. R. P. Todorov and Kh. G. Khristov, *Metal Science and Heat Treat.* **46**, 49-53 (2004).
6. G. F. da Silva, S. M. Tavares, J. M. Pardal , M. R. Silva, H. F. G. de Abreu, *J. Mater Sci.* **46**, 7737–7744 (2011).
7. V. Venegas, F. Caleyo, T. Baudin, J.M. Hallen, R. Penelle, *Corro Sci.* **5**, 1140–1145 (2009).
8. L. Chaowen, Y. Wang, Y. Chen, *J. Mater Sci.* **46**, 6424–6431 (2011).
9. F. Huang, X. G. Li, J. Liu, Y. M. Qu, J. Jia, C. W. Du, *J Mater. Sci.* **46**, 715–722 (2011).
10. C. Hurtado Noreña, P. Bruzzoni, *Mater. Scien and Eng.* **A 527**, 410–416 (2010)
11. J. Sojka, P. Váňová, P. Jonšta, L. Rytířová, Jerome, *Acta Metall. Slova*, **12**, 462 – 468 (2006).
12. M. Abdur Razzak. *Bull. Mater. Sci.* **34**, 1439–1445 (2011).
13. B.S. Motagi, Ramesh Bhosle. *Intern. Journ. of Engin. Res. and Devel.* **2**, 07-13 (2012).

Mater. Res. Soc. Symp. Proc. Vol. 1481 © 2012 Materials Research Society
DOI: 10.1557/opl.2012.1634

Tension Tests Behavior of API 5L X60 Pipeline Steel in a Simulated Soil Solution to Evaluate SCC Susceptibility

A. Contreras[1*], S. L. Hernández[1], R. Galvan-Martinez[2], and O. Vega-Becerra[3]

[1] Instituto Mexicano del Petroleó, Eje Central Lázaro Cárdenas Norte 152, Col. San Bartolo Atepehuacan, C.P. 07730, México.
*Email: acontrer@imp.mx
[2] Unidad Anticorrosión, Instituto de Ingeniería, Universidad Veracruzana, Ave. S.S Juan Pablo II S/N, Ciudad Universitaria, Fracc. Costa Verde, Veracruz, C.P. 94294, México.
[3] Centro de Investigación en Materiales Avanzados, S.C. Unidad Monterrey Alianza Norte 202. Parque de Investigación e Innovación Tecnológica. Apodaca, Nuevo León, C.P. 66600, México.

ABSTRACT

In this work slow strain rate tests (SSRT) were used for the evaluation of API 5L X60 in contact with a simulated soil solution called NS4 in order to evaluate stress corrosion cracking (SCC) susceptibility. SSRT were carried out in NS4 solution at room temperature to simulate dilute ground water that has been found to be associated with SCC of low carbon steel pipelines. A strain rate of 1×10^{-6} sec^{-1} was used. According to the analysis of SSRT results, the X60 pipeline steel is highly resistant to SCC at the conditions studied. A combine fracture type it was observed: ductile and brittle with a transgranular appearance. Some pits close to the fracture zone were observed. The failure process and mechanism of X60 steel in NS4 solution are controlled by anodic dissolution and hydrogen embrittlement which was revealed with the internal cracks observed in the surface fracture. There is a relation between the strength of the steel and the SCC susceptibility, thus, increasing strength in the steel, the SCC susceptibility increases as a function of the pH solution used.

Keywords: corrosion, steel, fracture, microstructure, embrittlement.

INTRODUCTION

It has been only a few years since it was recognized that stress corrosion cracking (SCC) can occur in buried pipelines. SCC on the external surface of pipelines has occurred in several countries throughout the world [1-3]. The failures provoke by SCC generally are catastrophic and almost always cause human injuries and economical losses. SCC has been experienced in line pipes with a wide range of chemical compositions, strengths, grades, sizes and coatings [4]. Additionally, it was seen that SCC has occurred in a wide variety of soils.

Pipeline steels are known to be susceptible to two types of stress corrosion cracking: intergranular (high pH-SCC or classical) and transgranular SCC (near neutral pH or non-classical). The great majority of those failures are associated with intergranular cracking, although transgranular cracking also have been observed. Intergranular cracking is suggested to occur by a localized dissolution process in a carbonate-bicarbonate solution. For high pH SCC it is well accepted that the mechanism involves anodic dissolution for crack initiation and propagation. In contrast, it has been suggested that the low pH SCC is associated with the dissolution of the crack tip and sides, accompanied by the ingress of hydrogen into the steel [5-7]. The hydrogen concentration plays an important role in this type of cracking. Cracks propagate as a result of anodic dissolution in front of their tip in SCC process, due to the embrittlement of their tip by hydrogen based mechanism.

The SCC failures are due to the fracture of metallic materials when they are subjected to stress in a corrosive solution that can be acidic, neutral or basic. These failures are more likely in acidic media, and there are many studies on the effect of concentration, temperature, the stress in the metal, roughness and the microstructure of the material [8-12].

The SCC was studied an X70 steel in the presence of an acid soil solution [8]. This research describes that the presence of more positive potentials the SCC is caused by an anodic dissolution mechanism. While more negative potentials SCC is generated due to the evolution of hydrogen [13,14].

SSR testing was used to examine the interaction between deformation produced by the stress applied along the time and exposure of the steel to corrosive solution. Time to failure (TF), reduction in area (%RA), yielding strength (YS), strain (%ε), and ultimate tensile strength (UTS) generally are measured as parameters indicative of the susceptibility of the steel to environmental degradation.

This research work studied the microstructural, mechanical and environmental SCC behaviour of X60 pipeline steel in a soil solution in function of the pH using SSRT to evaluate the SCC susceptibility.

EXPERIMENTAL DETAILS

Material

The material used in this study was an API 5L X60. The steel was studied in the conditions of as received, and the samples were obtained from a pipe of 1066.8 mm (42 inches) of diameter with 12.7 mm (0.5 inches) wall thickness. Chemical composition of X60 steel studied is shown in Table I and the mechanical properties in Table II.

Table I. Chemical composition of the API 5L X60 (wt.%)

C	Mn	Si	P	S	Al	Nb	Cu	Cr	Ni	Mo	Ti	Fe
0.020	1.57	0.14	0.013	0.0020	0.046	0.095	0.30	0.26	0.17	0.05	0.014	Bal.

Table II. Mechanical properties of the API 5L X60 pipe steel.

YS (MPa)	UTS (MPa)	Elongation (%)	HV
467	566	40	201

Test Solution

The solution used to carry out the SSRT was called NS4. This solution has been used widely to simulate the electrolyte of the soil where SCC has been observed in underground pipelines [15-17]. The soil chemical composition shows that the principal electrolytes contained are variable proportions of carbonates, bicarbonates, chlorides and sulphates mainly.
The SSRT in the soil solution (NS4) were carried out at strain rate of 1×10^{-6} s^{-1} at room temperature. The pH of NS4 solution obtained is ranged between 8 and 8.5. NS4 solution with pH of 3, 8 and 10 (adjusted to pH 3 with HCl and pH 10 with NaOH) was used to perform the SSRT in order to evaluate the SCC susceptibility. Table III shows the chemical composition of the soil solution used.

Table III. Chemical composition of the NS4 solution (gr/l).

NaHCO$_3$	CaCl$_2$. 2H$_2$O	MgSO$_4$. 7H$_2$O	KCl
0.483	0.181	0.131	0.122

Slow strain rate tests (SSRT)

In recent years the SSRT has become widely used and accepted for SCC evaluations of low carbon steels, stainless steel and others alloys [18]. SSR technique involves the slow straining of a specimen of the steel of interest in a solution in which will be in service. SSRT and SCC assessment were carried out according to requirements of NACE TM-0198 and ASTM G-129 [19,20]. Cylindrical tensile specimens with a gauge length 25.4 mm (1 inch) and 3.81 mm (0.150 inches) gauge diameter were used. A glass autoclave to carry out the SSRT in the soil solution was used. A schematic representation of the autoclave with the tensile specimen used is shown in Figure 1(a). The SSRT were performed in a MCERT (Mobile Constant Extension Rate Tests) machine containing the autoclave as is shown in figure 1(b).

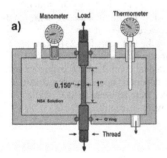

Figure 1. (a) Glass autoclave used to perform the SSRT with the NS4 solution and (b) MCERT machine used to carry out the SSRT.

After SSRT was completed, the surface fracture were immediately removed from the soil solution and cleaned to carry out the Scanning Electron Microscope (SEM) observations.

Fractographic analysis

After perform the SSRT, characterization of the fracture surface were performed in a XL-30 ESEM Philips Environment Scanning Electron Microscope (ESEM), in order to study the SCC susceptibility. The surface fractures were analyzed in order to evaluate the reduction in area, as well as to determine the type of fracture and its features. Additionally, some longitudinal gauge sections were observed in order to determine if cracks are presents in the samples, or if some evidence of pitting corrosion give the origin to the failure.

Assessment of the SCC susceptibility

The SCC susceptibility has been evaluated using SSRT in air (as an inert environment) and in a NS4 solution at room temperature and pH values of 3, 8 and 10 for the soil solution. The degree of susceptibility to SCC is generally assessed through observation of the differences in the behavior of the mechanical properties of the material in tests conducted in a specific environment (in this case the soil solution) from that obtained from tests conducted in the controlled environment (air). The assessment was performed for change in time to failure ratio (TFR), reduction in area ratio (RAR), strain ratio (εR) and elongation plastic ratio (EPR). Additionally, SEM observations for visual indications of cracks were carried out. In order to determine if the steel is susceptible to SCC, the combination of these methods according to requirements of NACE TM-0198 and ASTM G-129 were used [19,20].

RESULTS AND DISCCUSION

Microstructure

Figure 2 shows the microstructure of the API X60 steel used in this study. The structure consists of dark areas of pearlite and the light areas formed by ferrite. In addition, some bands of pearlite aligned according to the direction of deformation were observed.

Metallurgical factors affecting SCC has been studied in the last years [21-23]. Many researchers have reported that uniform microstructures are favorable for suppressing SCC. Low carbon steels produced in a process such as thermo-mechanical controlled processing are less susceptible to SCC. The difference of SCC susceptibility depended upon microstructures. Uniform microstructures, such as bainite or bainitic ferrite, were resistant to SCC, while inhomogeneous microstructures, such as ferrite-pearlite, were susceptible to SCC.

Figure 2. Typical microstructure of the API X60 pipeline steel.

Slow strain rate tests (SSRT)

Figure 3 shows the stress-strain profiles obtained from the SSR tests for the X60 steel tested at room temperature in air and NS4 solution at different pH values in order to assess the SCC susceptibility. It is clear that solution with pH 3 and 10 have an effect on the mechanical properties, which in turn will be observed in the SCC assessment.

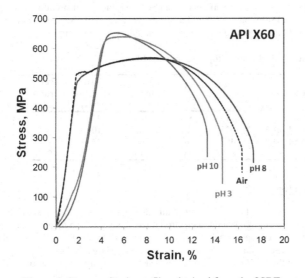

Figure 3. Stress vs Strain profiles obtained from the SSRT.

Assessment of the SCC susceptibility

The susceptibility to SCC was expressed in terms of the time to failure ratio (TFR), reduction in area ratio (RAR), strain ratio (εR) and elongation plastic ratio (EPR). Table IV shows the results of SCC assessment for the X60 steel for the different test conditions. The results indicate that X60 pipeline steel is highly resistant to SCC under the studied conditions. Ratios in the range of 0.8-1.0 normally means high resistance to environment assisted cracking (EAC), whereas low values (i.e.<0.5) show high susceptibility.

Table IV. Summary of results of SCC assessment for the X60 steel for the different test conditions obtained from the SSRT.

Condition	TF(h)	TFR	%RA	RAR	%ε	εR	%EP	EPR
Air	47.41		80.6		16.33		12.90	
NS4-pH 3	42.37	0.89	81.8	1.01	14.63	0.89	12.42	0.96
NS4-pH 8	50.67	1.06	80.3	0.99	17.33	1.06	16.25	1.26
NS4-pH 10	38.50	0.81	73.3	0.90	13.34	0.81	10.93	0.84

From the analysis of results showed in Table IV it is clear that X60 pipeline steel is highly resistant to SCC under the studied conditions, however the most susceptible condition was observed at pH 10, followed by tests performed at pH3.

Table V shows a summary of the mechanical properties obtained from SSRT relate to the SCC susceptibility. The data in Table V clearly show that SCC resistance of pipeline steels decreases, generally with increasing the yield strength and ultimate tensile strength for the same steel and microstructure. However, the correlation between SCC resistance and strength depend of the microstructure and soil solution used. Similar results were reported elsewhere [24] for an API X52 pipeline steel studied with different levels of cathodic protection applied.

Table V. Mechanical properties obtained from the SSRT related to SCC.

Condition	YS (MPa)	UTS (MPa)	Stress rupture (MPa)	Elastic Modulus (Mpa)	Average ratio
Air	513	567	268	245	
NS4-pH 3	625	638	301	106	0.93
NS4-pH 8	492	565	260	243	1.09
NS4-pH 10	638	651	315	105	0.84

Fractographic analysis

Complementary to assess the SCC susceptibility through TFR, RAR, εR, and EPR, the samples were observed through scanning electron microscopy at low magnification in order to observe if there was a secondary crack in the gauge section of the samples, which will be indicative of SCC. No secondary cracks were observed as is shown in figure 4a. Experimental data (D_f) to calculate the RAR were obtained from SEM micrographs as is shown in figure 4b.The primary fracture showed a combine ductile and brittle type of surface fracture as is shown in figure 4b. At higher magnification of the ductile zone is presented in figure 4c. Ductile fractures are characterized by extensive plastic (permanent) deformation of the material and the neck formation on the gage section before to material failure (figure 4a).The presence of some dimples and microvoids are evident which is a feature of the ductile fracture. A transgranular type of fracture was observed as is shown in figure 4d, overall for samples tested in NS4 solution with pH 3.

Secondary cracking in the gage section of the specimen is generally associated with conditions that promote SCC in the material. The analysis of ratios (TFR, RAR, EPR, etc) in combination with the SEM observation to find secondary cracks is very important to the interpretation of SSRT results.

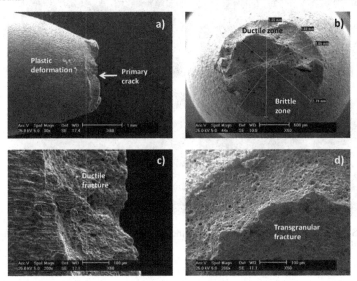

Figure 4. SEM images of fracture surfaces obtained from SSRT for specimens tested in NS4 solution with pH 8 at room temperature.

Most of the samples presented a great quantity of globular inclusions as is shown in the image of figure 5a. For the samples tested in NS4 solution the surface fracture showed some microvoids which were originated in some cases by the presence of some inclusions as is observed in figure 5a. These voids grow until adjacent voids connect, coalescing into larger voids as can be seen in figure 5b. Final failure of the material occurs when the voids grow larger and connect together. The microanalysis (EDS) performed in the inclusions revealed the presence of Mn, Al, Ca, S and C. Some of them are Fe_3C and MnS mainly. According to Lu *et al* [25] the inclusions acted as the crack initiation sites. Therefore, it is expected that SCC resistance of pipeline steel can be improved by reducing the inclusions content in the steel. For these specimens some internal cracks were observed as is shown in figure 5b, which are indicative the process to fail the material was through hydrogen induce cracking (HIC). Samples tested at pH 3 and pH 10 presented some pits close to the primary crack (surface fracture) as is shown in figure 5c and figure 5d. Since no passive film is formed on X60 steel in the simulated soil solution, the failure mechanism for the X60/NS4 system cannot be explained by the passive film rupture. Hence, hydrogen embrittlement (HE) and/or anodic dissolution (AD) are likely the mechanism of failure for X60 steel.

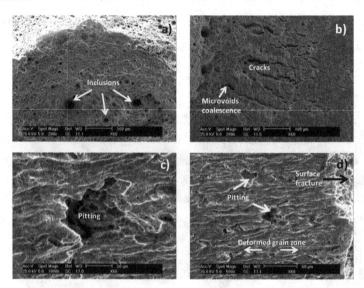

Figure 5. SEM images of fracture surfaces obtained from SSRT for specimens tested in NS4 solution with pH 8 at room temperature.

Some causes of SCC damage on pipeline steels are attributed to the crack initiation by hydrogen ingress. When the hydrogen is trapped in the defects such as microvoids and inclusions, the lattice is dilated and the interatomic cohesion decreased, thus alteration in the atomic microstructure enhances the presence of internal cracks.

Al-Mansour et al [26] studied the sulphide stress cracking (SSC) resistance of API X100 high strength steel; they suggested that SSC susceptibility was caused by the high corrosion rate which formed corrosion pits that acted as crack initiation sites on the metal surface. Corrosion pits are one of the main potential sites for surface crack nucleation. The stress concentration in the pit and the corrosive environment in addition to the mechanical characteristics of the material will give rise to crack nucleation and subsequent propagation.

In addition, minimizing the phase boundaries in the steel will decrease corrosion rates which in turn will minimize the pitting and reduce the intensity of hydrogen propagation into the steel. Another way to improve SSC resistance is to better control non-metallic inclusions.

Liang et al [27] studied the SCC susceptibility of X80 steel under applied cathodic potentials in a simulated soil solution using SSRT. They found that pits are an important factor in the initiation of cracks. Pitting in a current-neutral pH environment preferentially occurs at surfaces with the highest residual stresses as a result of long-range galvanic reactions (stress cell) [28]. Cracking, however, can only develop at the bottoms of pits located in regions with a moderate tensile residual stress at the surface and a low rate of reduction in tensile residual stress in the depth direction.

CONCLUSIONS

API X60 pipeline steel was evaluated through SSRT in order to assess the susceptibility to SCC in a soil solution (NS4) at pH values of 3, 8 and 10, a room temperature. The assessment results indicate that X60 pipeline steel was not susceptible to SCC at the conditions studied. Specimens tested in air exhibited a ductile type of fracture. Whereas, specimens tested in the soil solution showed a combination of both ductile and brittle fracture. A common feature of these fractures was quasi-cleavage fracture mixed with microvoid coalescence. Some of these fractures showed internal cracks due to hydrogen ingress into the steel. Some pits close to surface fracture were observed. Since no passive film is formed on X60 steel in the simulated soil solution, the most probable failure mechanism for the X60/NS4 system cannot be explained by the passive film rupture. Therefore, hydrogen embrittlement and/or anodic dissolution are likely the mechanism of failure for X60 steel.

REFERENCES

1. R. N. Parkins, *Corrosion*, **50**, 394 (1994).

2. M. Elboujdaini, Y.Z. Wang, R.W. Revie, International Pipeline Conference (IPC) ASME, 967 (2000).

3. J. A. Beavers, B. A. Harle, *Journal of Offshore Mechanics and Artic Eng.* **123**, 147 (2001).

4. B. N. Leis and R. J. Eiber, *Proceedings, first International Business Conference on Onshore Pipelines*, Berlin, December (1997).

5. B. Y. Fang, R. L. Eadie, W. X. Chen and M. Elboujdaini, *Corrosion Engineering, Science and Technology*, **45**, 302 (2010).

6. B.Y. Fang, A. Atrens, J. Q. Wang, E.H. Han, Z.Y. Zhu and W. Ke, *Journal of Materials Science*, **38**, 127 (2003).

7. X. Y. Zhang, S.B. Lambert, R. Sutherby and A. Plumtree, *Corrosion*, **55**, 297 (1999).

8. Z.Y. Liu, X.G. Li, C.W. Du, G.L. Zhai, Y.F. Cheng, *Corrosion Sci.*, **50**, 2251 (2008).

9. R. W. Bosch, *Corrosion Sci.*, **47**, 125 (2005).

10. X. Lou, P. M. Singh, *Electrochimica Acta*, **56**, 1835 (2011).

11. P. Kentish, *Corrosion Sci.*, **49**, 2521 (2007).

12. T. Shoji, Z. Lu, H. Murakami, *Corrosion Sci.*, **52**, 769 (2010).

13. S. Ramamurthy, W.M.L. Lau, A. Atrens, *Corrosion Sci.*, **53**, 2419 (2011).

14. Y. F. Cheng, *Electrochimica Acta*, **52**, 2661 (2007).

15. X. C. Li, R.L. Eadie and J.L. Luo, *Corrosion Engineering, Science and Technology*, **43**, 297 (2008).

16. A. Benmoussat and M. Hadjel, *Journal of Corrosion Science and Engineering*, **7**, 1 (2005).

17. L. J. Qiao, J. L. Luo, *Journal of Materials Science Letters*, **16**, 516 (1997).

18. R. D. Kane, C.J.B.M. Joia, A.L.L.T. Small and J.A.C. Ponciano, *Materials Performance*, **71** (1997).

19. NACE TM-0198 Slow Strain Rate Test Method for Screening Corrosion-Resistant Alloys (CRAs) for Stress Corrosion Cracking in Sour Oilfield Service, 1-21, (2004).

20. ASTM G-129, Slow strain rate testing to evaluate the susceptibility of metallic materials to environmentally assisted cracking, 1-7, (2006).

21. H. Asahi, T. Kushida, M. Kimura, H. Fukai, and S. Okano, *Corrosion*, **55**, 644 (1999).

22. M. Sawamura, H. Asahi, T. Omura, H. Kishikawa, N. Ishikawa and M. Kimura, Corrosion NACE Conference & Expo, Paper 11286 (2011).

23. T. Kushida, K. Nose, H. Asahi, M. Kimura, Y. Yamane, S. Endo, Corrosion NACE Conference & Expo, Paper 01213 (2001).

24. A. Contreras, S. L. Hernández, R. Orozco-Cruz, R. Galvan-Martínez, *Materials & Design*, **35**, 281(2012).

25. B. T. Lu and J. L. Luo, *Corrosion*, **62**, 723 (2006).

26. M. Al-Mansour, A. M. Alfantazi, M. El-boujdaini, *Materials & Design*, **30**, 4088 (2009).

27. P. Liang, X. Li, C. Du, X. Chen, *Materials & Design*, **30**, 1712 (2009).

28. W. Chen, G. Van Boven, R. Rogge, *Acta Materialia*, **55**, 43 (2007).

Characterization of Materials for Industrial Applications

Mater. Res. Soc. Symp. Proc. Vol. 1481 © 2012 Materials Research Society
DOI: 10.1557/opl.2012.1635

Vacuum Foaming of Aluminum Scrap

J. A. Garabito[1*], H. Granados[1], V. H. López[1], A. R. Kennedy[2], J. E. Bedolla[1]

[1] Instituto de Investigaciones Metalúrgicas, Universidad Michoacana de San Nicolás de Hidalgo, Edificio "U" Ciudad Universitaria, Morelia, Mich., México.
[2] Faculty of Engineering, University of Nottingham, University Park, NG72RD, UK.
E-mail: garabo_coria@hotmail.com

ABSTRACT

In this study, scrap from the automotive industry was used to produce aluminium foams under vacuum. Chips of an aluminium alloy LM26 were melted and 1wt. % of Mg was added for creating a viscous casting with uniform distribution of oxides. An ingot was obtained of this alloy after casting and solidification. Trials for foaming this alloy were performed by re-melting pieces under vacuum at different temperatures. A window in the vacuum chamber allowed observation of the foaming and collapse of the porous structures was observed during cooling. Characterization of the aluminum foams revealed different levels of expansion, porous structures and degrees of drainage. The best foams were obtained at 680 °C with a density of 0.78 g/cm^3. This technique appears to be a feasible low cost route for producing Al foams based on scrap material.

Keywords: Al, Casting, Foam, Oxide, Porosity.

INTRODUCTION

The potential applications of aluminum foams, in virtue of their unique combination of physical and mechanical properties, have been envisaged in the area of the automotive industry, light weight construction materials, silencers, flame arresters, heaters and heat exchangers, catalysts, electrochemical applications, military armour vehicles and aircraft [1].

However, an inherent drawback of this kind of materials is given by their high cost, which is basically due to the production process use to produce cellular metals. Fabricating aluminium foams entirely from scrap is attractive owing to the cost savings offered by a low-cost matrix and the potential for eliminating expensive and embrittling foam-stabilising additives. For example, maintaining the Fe content in the lower levels and avoiding long periods of cooling are key aspects for preventing the formation of β-type Al-Fe-Si precipitates in a recycled Al-Si-Mg alloy [2]. Re-melted Al scrap in the form of used beverage cans, has been used [3], in conjunction with TiH$_2$ blowing agent, to produce Al foams. It is reported that the scrap, via the oxides introduced from their surfaces, acts as a viscosity thickener, aiding foaming [3]. Highly distorted cell structures were, however, observed and attributed to the inhomogeneous distributions of the introduced oxides. It was concluded that the agglomerated oxide particles introduced could not act as a foam stabiliser. More recently, Alcoa reported the development of a commercial Al foam derived from scrap material made by a liquid route using CaCO$_3$ as a foaming agent [4]. In order to obtain aluminum foams with close porosity, the present work is based in the selection of an economical production process along with the use of low price raw materials (recycled aluminum). The aim of this study is to undertake a preliminary investigation into the expansion, structure and stability of foams made from 100% Al alloy scrap produced via a liquid route where foaming of the melt is induced by vacuum.

EXPERIMENTAL PROCEDURE

The scrap used was swarf in the form of mm-sized chips, as shown in Figure 1, a by-product of the machining of an LM26 alloy commonly used for manufacturing automotive castings. This material is the same that was employed in reference [5]. The chemical composition of the alloy, as measured by optical emission spectroscopy (OES), is given aside Figure 1. The approximate melting range for the alloy is 530-580°C according to a differential scanning calorimetry thermogram.

Processing of the scrap into ingots was performed by charging a crucible (preheated to 750 °C) with scrap. After melting of the scrap (which took roughly 30 min) the charge needed to be vigorously stirred to merge the individual molten chips. Owing to the poor packing of the swarf, it was necessary to add more material to produce a full charge and then 1 wt.% Mg was incorporated into the melt via an Al-Mg master alloy. The melt was kept at temperature for 4 h to "condition" the melt. Finally, the molten metal was cast into a steel mould. Samples were taken for microstructural characterization.

Element	wt. %
Si	10.50
Cu	1.60
Fe	1.20
Mn	0.29
Mg	0.13
Zn	1.10
Al	balance

Figure 1. Photograph of the swarfs and chemical composition.

The Al ingot was sectioned into small pieces of ~20 g. Trials for foaming the LM26-1wt% Mg alloy were performed by re-melting pieces of the alloy in a muffle furnace at 660, 680, 700, 720 and 740°C. Subsequently, liquid samples were introduced into a stainless steel chamber where vacuum was applied enabling foaming. The reduced pressure caused the dissolved gases to expand and enhanced volatilization of Zn and Mg, creating a porous structure. The vacuum chamber had a window that allowed observation of the evolution of the foaming process. Prior to the foaming experiments, the mass of the samples was recorded. Expansion of the foams was quantified according to the equation

$$\% \, Expansion = \left(\frac{Vf - Vi}{Vi}\right) x \, 100 \quad (1)$$

Where, V_i is the initial volume and V_f is the final volume. The density of the foams was measured from the mass and dimensions of the samples, taking into account foaming conditions and considering the diameter and height of the foamed sample. The aluminum foams were cut in half along the axis of expansion using a hacksaw. The foam sections were scanned and the resulting images were analyzed with software facilities.

RESULTS AND DISCUSSION

The purpose of adding Mg into the melt and leaving it for long time at temperature was aiming to generate a network of oxides within the molten metal and a viscous melt. Figure 2 shows the microstructure of the ingot obtained from the scrap. The images reveal a quite different microstructure between top and bottom of the ingot as a result of the different cooling rates. Beyond the grain size, content and size of precipitates, the x-ray dot maps did not disclose the presence of a network of oxides, instead, only a few scattered clusters of oxides may be pointed allocated in the grain boundaries along with second phases. Thus, the amount of Mg added and holding at temperature during 4 hours failed to produce such a network of oxides as compared to the procedure used in [5] where the swarf was pre-treated at 500°C in air for 24 h.

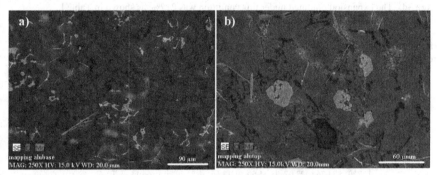

Figure 2. Mg and O_2 X-ray dot maps at a) bottom and b) top of the ingot made of the re-melted scrap with 1wt.% Mg addition.

Figure 3 presents the porous structures obtained after the vacuum foaming trials. Observations during foaming revealed that the alloy expanded quite well initially, but after reaching a maximum in the expansion, the structure steadily collapsed in the vacuum chamber. Generally speaking, the samples exhibit a coarse and quite irregular porous structure with considerable drained material at the base of the foam. This accumulation of liquid metal indicates that extensive, gravity-driven drainage of liquid metal occurs through the cell structure during cooling of the foam before freezing. According to Kumar [5] Mg addition has two roles; reduce the bulk viscosity of the liquid and increase it at the surface by forming oxides that aid the stabilization of the foam structure. However, the interplay between temperature and reduced pressure seems to dictate in the first instance the proper condition for foaming. A superheat greater than 100°C leads to collapsed structures, which means that rapid reduction of the pressure in the systems is immediately reflected on gas evolution within the melt and thereby on expansion. At this point, retention of the porous structure relies on rapid cooling and freezing [6-8].

85

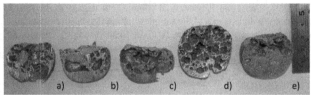

Figure 3. Macrographs of LM26-1%Mg porous structures foamed under vacuum at a) 740°C, b) 720°C, c) 700°C, d) 680°C and e) 660°C

In atmospheric Al casting conditions, mostly, micro-pores appear during solidification due to the dramatic reduction of hydrogen solubility from the liquid to the solid phase (If care is not taken in preventing sources of contamination). In this study, gas evolution occurred when the liquid samples were in liquid state, under this circumstance the presence of solids within the melt have been found decisive for pore nucleation [7, 9]. Thus, the scattered oxides observed in the ingot, although were not effective in stabilizing the foam structure; they enabled extensive pore nucleation and foaming of the melt.

In terms of expansion and porous structure, the sample foamed at 680°C was outstanding (Figure 3d). The expansion obtained under this condition was 217%, as shown in Table I, with a density of 0.78 g/cm³. However, the shape of the foam is distorted and there is a dense metal layer at the base of the foam. A statistical analysis of the characteristics of the foam was performed and the details are given in Table II regarding shape factor, Feret diameter and pores area.

Table I. Estimation of the expansion of the foams at different temperatures.

T (°C)	W(g)	H_f(mm)	H_i (mm)	V_i (mL)	V_f (mL)	% Expansion
740	23.95	28.97	12.74	10.51	29.85	184.12
720	34.26	23.07	15.50	13.66	22.70	66.11
700	33.86	24.18	15.39	13.54	24.04	77.56
680	33.39	39.39	15.27	13.40	42.50	217.25
660	38.53	29.83	16.64	14.97	30.90	106.42

Table II. Statistics of the foam produced at 680°C.

	Shape factor	D Feret (mm)	Area (mm^2)
Average	0.77	0.80	1.77
Std. dev.	0.47	1.27	6.5
Pores analyzed	273	273	273

The calculated values of expansion and density of aluminum foams are not the only parameters that will determine the quality of the foam. Also, the internal structure of the foams is necessary to assess [10]. The pore density parameter (pores/cm^2) indicates the number of pores present per unit area on the section cut in the direction normal to expansion. So the calculated value is 21 pores/cm^2. The number of pores present has a direct relationship with the density of the foam expansion; however, the morphology of the pores causes different structures with similar densities. Thus, the porosity should be characterized in terms of size and shape, once quantified these variables, the size distribution and shape definition of the relative pore constitute an important complement characterization method. The pore size is defined in terms of the Feret diameter, while the shape of the pores by the form factor.

Figure 4 shows frequency of pore classes analysis for the pore size distribution of the best foam produced. The frequency chart summarizes the percentage of pores in each category increasing by 0.67 millimeters. The histogram shows that the pores of the foamed by vacuum mainly fluctuates in the range of 0.13 to 0.75 mm, this pore size interval exhibit a more homogeneous distribution, so that 75% of the pores are in this sizes. Researchers [11-12] noted that pores of less than 0.7 mm are not suitable for impact absorption applications as small pores relate to thick wall thickness. Pores larger than 11 mm do not have structural functionality.

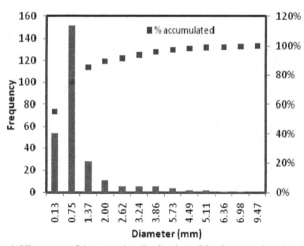

Figure 4. Histogram of the pore size distribution of the foam produced at 680°C.

CONCLUSIONS

The main outcome of this preliminary study is that it demonstrates to be possible obtaining at low cost aluminum foams from recycled material and without foaming agent by simply casting a melt under vacuum. Further work is necessary to achieve the physical properties of the melt that will prevent coalescence of pores, stabilize the foam structure and thereby avoid excessive drainage of molten metal.

REFERENCES

1. J. Banhart, *Prog. Mat. Sci.*, **46**, 559-632 (2001).
2. P. Fernández, L. J. Cruz and L.E. García-Cambronero, *Prospectiva*, **8**, 7-11 (2010).
3. W. Ha, S.K. Kim, H-H Jo and Kim Y-J, *Mat. Sci. Tech.*, **21** (2005).
4. http://www.alcoa.com/global/en/innovation/papers_patents/pdf/al_foam_story.pdf, in.
5. G. S. V. Kumar, K. Heim, F. Garcia-Moreno, J. Banhart and A. R. Kennedy, Busan (Korea) 2011.
6. H. Wiehler, C. Körner and R. F. Singer, *Adv. Eng. Mat.*, **10**, 171-178 (2008).
7. G. V. Kumar, M. Mukherjee, F. Garcia-Moreno and J. Banhart, *Metall. Mat.s Trans. A*, 1-8 (In press).
8. H. Nakajima, *Prog. Mat. Sci.*, **52**, 1091-1173 (2007).
9. C. C. Yang and H. Nakae, *J. Mat. Proc. Tech.*, **141**, 202-206 (2003).
10. B. Jianga, Z. Wanga and N. Zhao, *Scripta Mate.*, **56**, 169-172 (2007).
11. C. Korner, M. Arnold and R. F. Singer *Mat. Sci. Eng.*, 28-40 (2005).
12. C. Gregory and F. John, *J. Col. Inter. Sci.*, **127**, 222-238 (1989).

Mater. Res. Soc. Symp. Proc. Vol. 1481 © 2012 Materials Research Society
DOI: 10.1557/opl.2012.1636

Study of the Thermoluminescent Characteristics of Ceramic Roof Tiles Exposed to Beta Radiation

A. R. García-Haro[1,*], R. Bernal[2], C.Cruz-Vázquez[1], S.E. Burruel-Ibarra[1], V.R. Orante-Barrón and F. Brown[1]

[1]Departamento de Investigación en Polímeros y Materiales de la Universidad de Sonora, Apdo. Postal 130, Hermosillo, Sonora 83000 México.

[2]Departamento de Investigación en Física de la Universidad de Sonora, Apdo. Postal 5-088, Hermosillo, Sonora, 83000, México.

*E-mail: argh@gimmunison.com

ABSTRACT

In this work, thermoluminescence (TL) characteristics of roof tile ceramic samples previously exposed to beta radiation are reported for the very first time. TL measurements were carried out using powdered samples obtained by the the fine-grained method, with grain size ranged from 300 nm to 5 μm. Characteristic thermoluminescence glow curves showed a complex structure with a dosimetric maximum located at ~ 200 °C. TL response of roof tile samples increases as the radiation dose increases in the 25 Gy to 1.6 kGy range. One response showed a linear behaviour, with no evidence of saturation within the dose interval investigated. The entire TL glow curve exhibited a remarkable reusability during 10 consecutive irradiation-TL readout cycles. The total TL signal showed a very low fading and remained almost constant after 3 h of irradiation and the corresponding TL readout. TL dosimetry features of powdered roof tile place it as a promising material in retrospective dosimetry as well as in possible TL dating applications.

Keywords: ceramic, grain size, radiation effects, structure, Scanning electron microscopy (SEM)

INTRODUCTION

Several natural ceramic materials have been used as building materials; for example, tile, granite, marble, tiles, bricks, etc. and they exhibit thermoluminescence (TL) [1].
Thermoluminescence (TL) is a widely used and reliable technique in the field of radiation dosimetry. In thermoluminescence dosimetry the relationship between the TL signal of the material and the absorbed dose to be measured must be determined by an appropriate calibration. The areas of application aimed at monitoring the radiation dose absorbed are personnel dosimetry, environmental dosimetry, clinical dosimetry and retrospective dosimetry [2]. In the case of retrospective dosimetry, heated dosimeters of known exposure times and of well-known age as well are used, for instance, bricks and/or tiles, in order to discriminate doses originating from sources different than the ones coming from natural environment [3]. Retrospective dosimetry has its main application in accidental dosimetry, in which it is feasible to accomplish the determination of absorbed dose due to accidentally contaminated areas, above the normal background radiation. Accidental dosimetry also includes the determination of radiation doses

during events such as nuclear weapons explosions, nuclear reactor accidents or other incidences of unintended radiation release [1,3,4]. Some materials proposed for the mentioned application are quartz, NaCl, ceramic materials, calcium, silicate bricks, red bricks, and other building materials which are laid in homes, offices or schools, hence, they can act as thermoluminescence dosimetric materials in case of radiological accident [1,3,4,5,6,7,8]. A few of these materials are used in TL dating as well, an useful technique to determine the dose absorbed by natural materials resulting from exposure to naturally occurring radio-nuclides in the environment [9]. One of methods used in the preparation of samples for TL dating measurements is the fine-grained method, consisting in the selection of particles by their grain size based on their settling time in acetone [9,10].

In this work, TL dosimetry properties of roof tile powder obtained by the fine-grained method are reported as a first stage for possible TL dating and TL retrospective dosimetry applications in the near future, as well as to evaluate its potential use as a thermoluminescent dosimeter.

EXPERIMENTAL PROCEDURE

Roof tile samples were obtained by the fine-grained method described above. Initially, roof tile samples were ground inside a Planetary Mono Mill Pulverisette 6 ball mill. Milled roof tile samples were immersed in acetone inside a test tube; the mixture formed was shaken and allowed to settle during 2 min, leaving particles of less than 10 micron diameter still suspended in acetone. The suspension was poured off into another test tube, where remaining particles were allowed to settle during 20 min. The suspension was then discarded leaving the fraction of interest in the sediment. This fine-grain fraction was dried in order to evaporate the excess of acetone and used for characterization.

A Risø TL/OSL model TL/OSL-DA-20 unit equipped with a ^{90}Sr beta radiation source was used to perform beta irradiations, TL measurements and the TL dosimetry characterization. Aliquots of ~ 1 mg of roof tile powder samples placed in stainless steel cups were used for the readouts. All irradiations were accomplished using a 5 Gy/min dose rate at room temperature (22 °C). The TL readouts were carried out under N_2 atmosphere using a heating rate of 5 °C/s. Scanning electron microscopy (SEM) images were obtained using a JEOL JSM 5410LV model scanning electron microscope equipped with energy dispersive X-ray microanalysis system (INCA EDS detector, Oxford Instruments), operating at 25 kV.

RESULTS AND DISCUSSION

Results on the semi-quantitative analysis carried out by EDS show that chemical composition of the roof tiles samples obtained in this work is: Si, O, Fe, C, Ca, K, Al, Mg, Zn, Ti, Cu and Na, with the common composition of SiO_2; (49 %), Al_2O_3 (13 %), FeO (2 %), K_2O (2 %), Na_2O (1.7 %), MgO (1.9 %), CuO (1 %), ZnO (0.7 %), TiO_2 (0.3 %) and CaO (2%). These values were obtained directly from the X-ray microanalysis system of the EDS detector. An EDS spectrum corresponding to one of the samples is shown in figure 1.

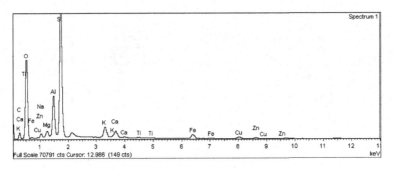

Figure 1. EDS spectrum of a roof tile sample.

Figure 2 shows the SEM image of a roof tile powder. The morphology consists of irregular round particles with sizes in the range from 300 nm to 5 μm, evidence that indicates fine-grain selection was accomplished.

Figure 3 shows the evolution of the TL glow curves from powdered roof tile samples exposed to beta radiation in the 0.025-1.6 kGy dose interval. As it can be observed, TL intensity increases as the dose increases, with no evidence of saturation within the entire dose interval studied. The maximum located at ~ 150 °C disappears for doses above 200 Gy. TL glow curves show a complex structure consisting of an intense maximum located at ~100 °C, and two less intense maxima located at ~ 150 and~ 200 °C. Regarding the glow peak located at ~ 200 °C, it can be considered as the dosimetric component of the entire TL glow curve, since it is well known that TL peaks located between 200 and 250 °C are suitable for dosimetry due to their stability under environmental conditions [2].

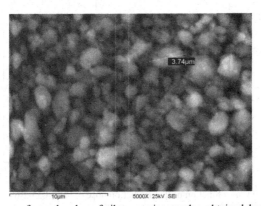

Figure 2. SEM image of powdered roof tile ceramic samples obtained by the fine-grained method.

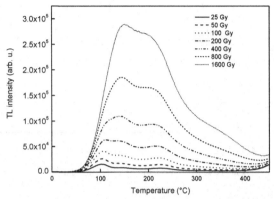

Figure 3. Thermoluminescence glow curves of roof tile powdered samples exposed to beta radiation in the 0.025-1.6 kGy dose interval.

Dose response is shown in figure 4. Each point within the figure represents the integrated TL of each glow curve from figure 3. Integrated TL as a function of the irradiation dose displayed a close linear behavior since the slope value obtained from the linear fit applied to the experimental data was 0.99939. The dose response of powdered roof tile can be useful as calibration curve which might be employed for retrospective dosimetry after a high-dose radiological incident.

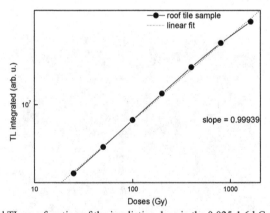

Figure 4. Integrated TL as a function of the irradiation dose in the 0.025-1.6 kGy interval.

The entire TL glow curve of roof-tile samples displayed a remarkable reproducibility according to the results obtained from 10 consecutive irradiation-readout cycles, as it can be observed in figure 5a. The normalized integrated-TL remained almost in a constant value with some fluctuations as shown in figure 5b. Reusability of a TL glow curve during irradiation-readout cycles is a desirable characteristic for any material to be considered suitable for TL dosimetry applications [2,11].

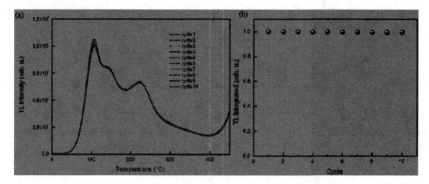

Figure 5. (a) Glow curves of roof tile samples obtained after 10 consecutive irradiation readout cycles and (b) Total TL signal as a function of cycle number. The samples were exposed to 100 Gy of beta radiation.

Figure 6 shows the integrated TL as a function of the time interval elapsed between irradiation of roof tile samples with 50 Gy of beta particles and the corresponding TL readout in order to observe the TL fading behavior of the material. The integrated TL faded down 15 % during the first 3 h, tending to a constant value for longer times. Further studies on TL fading of this ceramic material are needed since usefulness of any phosphor for retrospective dosimetry depends on an understanding of the signal fading characteristics, in order to allow corrections to the dose estimate after a radiological incident.

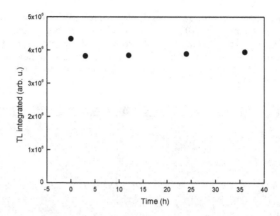

Figure 6. Integrated TL as a function of the time interval elapsed between the irradiation and the corresponding readout. The samples were exposed to 50 Gy of beta radiation.

CONCLUSIONS

In this work, thermoluminescence (TL) characteristics of roof tile ceramic samples previously exposed to beta radiation are reported for the very first time. According to the experimental results obtained from this investigation, such as linear dose response, low TL fading, and TL signal reusability, powdered roof tile obtained by a fine-grained method exhibits TL features suitable for high-dose retrospective dosimetry, useful in nuclear or radiological accidents. It is worth mentioning that TL signal reusability and stability of the studied material are appropriate characteristics for subsequent further studies on TL dating. The shape of the glow curves indicates that use of roof tile from local buildings as samples to evaluate dose within accident areas allows dose evaluation still many years after the accident. According to the experimental results obtained from this investigation such as linear dose response, low TL fading, and TL signal reusability, powdered roof tile obtained by a fine-grained method possesses suitable TL dosimetry features for high-dose retrospective dosimetry useful in nuclear or radiological accidents. It is worth mentioning that TL signal reusability and stability of the material studied are appropriate characteristics for subsequent further studies on TL dating.

REFERENCES

1. K. V. R. Murthy. IOP Conf. Series, *Materials Science and Engineering* **2** (2009), 012024O.
2. R. Chen and S. W. S. McKeever, *Theory of Thermoluminescence and Related Phenomena* (World Scientific, Singapore, 1997).
3. G.S. Polymeris, G. Kitis, N.G. Kiyak, I. Sfamba, B. Subedi and V. Pagonis, *Applied Radiation and Isotopes* **69**, 1255–1262 (2011).
4. I.K. Bailiff and V.B. Mikhailik, *Radiation Measurements* **38**, 91 – 99 (2004).
5. G. Espinosa, J.I. Golzarri, and P. Santiago, *Revista Mexicana de Física S* **54** (1), 17-21 (2008).
6. H.Y. Göksu, I.K. Bailiff and V.B. Mikhailik, *Radiation Measurements* **37**, 323–327 (2003).
7. I.K. Bailiff, *Radiation Measurements* **24**, 507–511 (1995).
8. I.K. Bailiff, *Radiation Measurements* **27**, 923–941 (1997).
9. O.B. Lian and D.J. Huntley in: Tracking Environmental Change Using Lake Sediments, Basin Analysis, Coring, and Chronological Techniques edited by W.M. Last and J.P. Smol, Vol. 1, Developments in Paleoenvironmental Research, chapter 12, Kluwer Academic Publishers (2002).
10. D.W. Zimmerman, *Archaeometry* **13**, 29 (1971).
11. H S.W.S. McKeever, *Thermoluminescence of Solids* (Cambridge University Press, Oxford, 1985).

Mater. Res. Soc. Symp. Proc. Vol. 1481 © 2012 Materials Research Society
DOI: 10.1557/opl.2012.1637

Evaluation of Resistance Spot Welding Conditions Using Experimental Design

D.Y. Medina[1*], R. Bermejo[1], R.T. Hernandez[1], I. Hernandez[1] and S. Orozco[2]

[1]DCBI, UAM-A, Av. Sn Pablo 180, 02200 México D.F., México.

[2]Fac. de Ciencias UNAM, Av. Universidad 3000, Col. Copilco el Bajo, México D.F., México.

* E-mail: dyolotzin@correo.azc.uam.mx

ABSTRACT

The breaking strength that can withstand solder is known as resistance welding, low resistance welding will cause a failure of the weld. This study optimized the effect of the main variables in a welding steel proceess on the mechanical propierties of the steel like the resistance welding. The main variables studied were electrode, machine, post induction, and dotted pressure. The factors that have the most influence in the resistance welding are the post induction, and the combination of post induction and the raw material.The statistical model used for the evaluation process was an analysis of variance (ANOVA) in a full crossing factorial design 2k with a second order of interaction.

Keywords: steel, strength, welding, soldering, simulation.

INTRODUCTION

The spot welding process is one of the most important methods in the manufacture of automotive assemblies which consist consists in a passing of a high electric current through two electrodes with tip. During the last few years, the requirements for safety and efficient materials in the design of the automotive body structure was increased and the spot welding process is an efficient response for joining process widely used for the fabrication of sheet metal assemblies which has excellent techno-economic benefits such as low cost, high speed and suitability for automation. Due to the large number of spot welds in a particular application, the process parameters need to be fine tuned. A tensile-shear strength testing is an important aspect of a weldability study, and the fracture point in the sample is knows at resistance spot welding (RSW).

Many studies were investigated the mathematical optimization of factors which interview in the response in the properties of materials for a welding process in a ferrous and non ferrous[1] materials, the studies were focused in the sample size[2-5] (width, overlap, thickness and length) and in the weld (quality, size and location)[1,6-10] but there are scarce literature about the influence of the spot welding process parameters in the RSW for steel [4,6]

In this study the optimal parameters as electrode, machine, post induction, dotted pressure, among others were researched. The analysis of variance (ANOVA) employed to obtain the effect of the process parameters on mechanical properties is an effect system to found a real solution [11].

EXPERIMENTAL PROCEDURE

The effects of the parameters were calculated with ANOVA in an experimental design using a matrix 2^x based on the factorial design, with a second order of interaction. The process parameters and levels were shown in Table I. An experimental plan for six parameters with three levels was organized in 2^6 matrix. The response of the testing was RSW as shown in figure 1.

The resistance welding tests were performed using an Instron machine. The charge used in the tensile test was 4000kN. And the samples have a diameter of 5mm and 10cm of length.

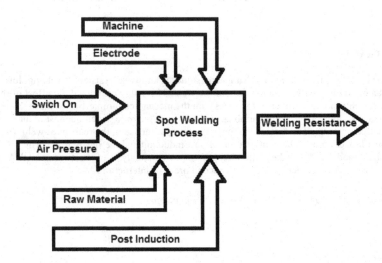

Figure 1. The parameters of the experiment and its response.

Table I. Experimental matrix 2^6.

Factor	Switch On (A)	Air Pressure (B)	Machine (C)	Electrode (D)	Raw Material (E)	Post Induction (F)
Low Level	All Machines	2 bar	1	beryllium	ABC	4 H
High Level	Spot Welder	8 bar	2	tungsten	S#B	6 H

RESULTS AND DISCUSSION

The experiments arrange with correspond resistance welding evaluation result, were shown in Table II. In each one of the probe performed one of the parameters was varied and the change of the resistance welding was tested.

Table II. Experimental Matrix with 64 probes.

Probe No	Switch on	Air pressure	Machine	Electrode	Raw Material	Post Induction	Resistance Welding Tested (MPa)
1	All	2	1	berilium	ABC	4	1400
2	Welder	2	1	berilium	ABC	4	1120
3	All	8	1	berilium	ABC	4	990
4	Welder	8	1	berilium	ABC	4	1300
5	All	2	2	berilium	ABC	4	1400
6	Welder	2	2	berilium	ABC	4	1300
7	All	8	2	berilium	ABC	4	1750
8	Welder	8	2	berilium	ABC	4	1380
9	All	2	1	tungsten	ABC	4	950
10	Welder	2	1	tungsten	ABC	4	1200
11	All	8	1	tungsten	ABC	4	900
12	Welder	8	1	tungsten	ABC	4	910
13	All	2	2	tungsten	ABC	4	1125
14	Welder	2	2	tungsten	ABC	4	1000
15	All	8	2	tungsten	ABC	4	900
16	Welder	8	2	tungsten	ABC	4	950
17	All	2	1	berilium	S#B	4	950
18	Welder	2	1	berilium	S#B	4	975
19	All	8	1	berilium	S#B	4	1100
20	Welder	8	1	berilium	S#B	4	1000
21	All	2	2	berilium	S#B	4	900
22	Welder	2	2	berilium	S#B	4	950
23	All	8	2	berilium	S#B	4	900
24	Welder	8	2	berilium	S#B	4	950
25	All	2	1	tungsten	S#B	4	1000
26	Welder	2	1	tungsten	S#B	4	1000
27	All	8	1	tungsten	S#B	4	1100
28	Welder	8	1	tungsten	S#B	4	1000
29	All	2	2	tungsten	S#B	4	1000
30	Welder	2	2	tungsten	S#B	4	1300
31	All	8	2	tungsten	S#B	4	1325
32	Welder	8	2	tungsten	S#B	4	1100
33	All	2	1	berilium	ABC	6	930

34	Welder	2	1	berilium	ABC	6	2000
35	All	8	1	berilium	ABC	6	2700
36	Welder	8	1	berilium	ABC	6	1920
37	All	2	2	berilium	ABC	6	1880
38	Welder	2	2	berilium	ABC	6	1600
39	All	8	2	berilium	ABC	6	1600
40	Welder	8	2	berilium	ABC	6	1460
41	All	2	1	tungsten	ABC	6	1460
42	Welder	2	1	tungsten	ABC	6	1300
43	All	8	1	tungsten	ABC	6	1100
44	Welder	8	1	tungsten	ABC	6	1000
45	All	2	2	tungsten	ABC	6	970
46	Welder	2	2	tungsten	ABC	6	600
47	All	8	2	tungsten	ABC	6	1000
48	Welder	8	2	tungsten	ABC	6	1100
49	All	2	1	berilium	S#B	6	440
50	Welder	2	1	berilium	S#B	6	1110
51	All	8	1	berilium	S#B	6	1600
52	Welder	8	1	berilium	S#B	6	1200
53	All	2	2	berilium	S#B	6	1400
54	Welder	2	2	berilium	S#B	6	1200
55	All	8	2	berilium	S#B	6	800
56	Welder	8	2	berilium	S#B	6	1400
57	All	2	1	tungsten	S#B	6	1400
58	Welder	2	1	tungsten	S#B	6	1400
59	All	8	1	tungsten	S#B	6	1400
60	Welder	8	1	tungsten	S#B	6	1400
61	All	2	2	tungsten	S#B	6	1400
62	Welder	2	2	tungsten	S#B	6	1400
63	All	8	2	tungsten	S#B	6	1380
64	Welder	8	2	tungsten	S#B	6	1400

   The result of ANOVA is shows in Figure 2 Standardized pareto chart for resistance spot welding (RSW). In this chart we can see that the most positive effect in the resistance welding is for the combination of parameters D (electrode) and E (raw material). And the most negative effect in the resistance welding is with the changing of the electrode from berilium to tungsten. The parameters B, C, and A (Air pressure, machine and swuitch on) not affect the RSW, also the second order combination of parameters, CF, BD, BC, BF, don't affect the RSW too. These results were analyzed with a 95% of confidence interval. The Confidence intervals are an expression of probability and are subject to the normal laws of probability. If several statistics are presented with confidence intervals, each calculated separately on the assumption of independence, that assumption must be honored or the calculations will be rendered invalid.
   The results of the interaction effects of the combination parameters show in Figure 3. The most interaction observed is the combination of effect D and E, its means that the effect of D and

E combination is bigger that each simple effect, in other words if the electrode is changed and the raw material is changed too, the RSW tested was increased significantly, but if only changed the electrode or the raw material the RSW tested was decreased.

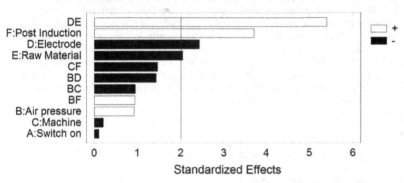

Figure 2. Standardized Pareto Chart for Resistance Spot Welding.

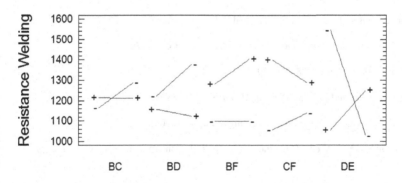

Figure 3. Interactions Chart for Resistance Spot Welding

CONCLUSIONS

The experimental design was an efficient tool for the optimization of the resistance welding effect in the steel, that improve the welding process, because found the significant effects that improve the resistance welding and the significant effects that retrogress the resistance welding too. In other hand the experimental design and the ANOVA, allows despise the factors which not have a significant effect. The experimental design and the ANOVA realized shown that the factors that had a significant effect in a RSW with a 95% of confidence interval were in first place, the positive effect of the combination D and E (electrode and raw material), continued for the positive effect of factor F (post induction) and in a low significance the negative influence of factor D (electrode) and the factor E (raw material), The ANOVA probe shown that the others factors don't have a significant effect. There before to increase the RSW is necessary to control both factors in the same way; electrode and raw materials, because if not the desirable effect of the electrode or the raw material in the RSW were exhibit. Other studies in order to increase the levels in these parameters are now evaluating.

REFERENCES

1. B. Ozcelik, Int. Commun. *Heat Mass Transfer,* **38**, 1067 (2011).

2. S. Aslanlar, *Mater Design,* **27**, 125 (2006).

3. M. Ouisse and S. Cogan, *Mechanical Systems and Signal Processing,* **24**, 1172 (2010).

4. H. Yang, Y. Zhang, X. Lai, and G. Chen, *Mater Design,* **29**, 1679 (2008).

5. Y. Zhang and D. Taylor, *Finite Elements Anal. Des.* **37**, 1013 (2001).

6. S. Aslanlar, A. Ogur, U. Ozsarac, E. Ilhan, and Z. Demir, *Mater Design,* **28**, 2 (2007).

7. S. Chae, K. Kwon, and T. Lee, *Finite Elements Anal. Des.* **38**, 277 (2002).

8. N. Kahraman, *Mater Design,* **28**, 420 (2007).

9. X. C. Li, D. Farson, and R. Richardson, *J. Manuf. Syst.* **19**, 383 (2001).

10. P. Salvini, F. Vivio, and V. Vullo, *Int. J. Fatigue* **22**, 645 (2000).

11. P. B. Harrington, N.E. Vieira, J. Espinoza, J.K. Nien, R. Romero, A.L. Yergey. *Anal. Chim. Acta,* 544 (2005).

Characterization of Materials Used in Coatings and Thin Films

Mater. Res. Soc. Symp. Proc. Vol. 1481 © 2012 Materials Research Society
DOI: 10.1557/opl.2012.1638

Diffusion of Hard Coatings on Ductile Cast Iron

N. López Perrusquia[1*], M. Antonio Doñu Ruiz[1], E. Y. Vargas Oliva[2], V. Cortez Suarez[3]

[1]Instituto Politécnico Nacional, SEPI-ESIME, U. P. Adolfo López Mateos, Zacatenco, 07738, México.
[2]Universidad Politécnica del Valle de México, Grupo Ciencia e Ingeniería de Materiales, 54910, Tultitlán, Estado de México
[3]Universidad Autónoma Metropolitana Unidad Azcapotzalco, México.
*E-mail: nlopezp0803@ipn.mx

ABSTRACT

This work estimate the growth kinetics of Fe_2B coatings created on surface nodular cast iron ASTM A-536 class 80-56-06. The Fe_2B coatings were formed by power packaging boriding process, considering three temperatures and exposure times different treatment. The hard coatings were evaluated through X-ray diffraction (XRD) and scanning electron microscopy (SEM). The model of diffusion employs the mass balance equation at the (Fe_2B/substrate) interface to evaluate the boron diffusion coefficient in the Fe_2B coating D_{Fe2B}, an expression of the parabolic growth constant, the instantaneous velocity of the Fe_2B/substrate interface, and the weight gain in the boriding sample were establish as a function of the parameter $\varepsilon(T)$ and $\eta(T)$, dependents of boriding process in function of the temperature related and the time of boriding $t_0(T)$, respectively in the Fe_2B coating. Model validation was extended considering the treatment of 1273 and 1123 K for 10 h respectively, obtaining a good correlation with experimental data.

Keywords: diffusion, kinetics, x-ray diffraction, scanning electron microscopy (SEM), packaging.

INTRODUCTION

Boronizing is technically well developed and widely used in industry to produce extremely hard and wear resistant surface layer on metallic substrate. This treatment is similar to other surface hardening treatments like carburization and nitriding in respect of physical and chemical characteristics [1-2]. It is successfully applied to all ferrous materials, nickel alloys, titanium alloys, and sintered carbides. This thermochemical treatment, carried out at temperatures in the range 800–1100°C for periods varying between 1 and 12 h, using gases or powders as boronizing media, gives rise to hard coatings constituted by an inner layer of Fe_2B (Boron rich 8.83 wt. % B) and an outer layer of FeB (Boron rich 16.23 wt. % B) and thickness in the range of 40–270 μm [3]. Ductile cast iron is the one in which the graphite is present as tiny spheres. The relatively high strength and toughness of ductile iron give it some advantages over other cast iron and steel types in many structural applications and industrial applications requiring wear and corrosion resistance; components include drive shafts, camshafts, pulleys, machine slide-ways, tanks, weapons and part for agricultural machinery.

To control the boronizing processes, knowledge of kinetic parameters is essential. The recent work respecting the modeling growth kinetic of Fe_2B coating has been developed for the establishment of the variables that affect the boronizing process. It is very important to establish the variables that affect the boronizing kinetics to control automated procedures and to obtain desirable properties [4]. The growth kinetics of Fe_2B coating formed on the surface of ductile iron ASTM A-536 were estimated for XRD and EDS microanalyses inform about the distribution of the alloying elements in the boride layer. Also, a relationship is proposed describing the evolution of the parabolic growth constant, the weight-gain and the instantaneous velocity of the Fe_2B/substrate interface, as a function of $\varepsilon(T)$ and $\eta(T)$ parameters, based on both boride incubation time and process temperature The results were found to be in a good agreement with the experimental conditions used in this work.

Diffusion Model

The model considers the substrate saturated with boron atoms, the assumptions were taken into account (see [5-6] further details) and coating with developing discontinuous globular graphite at the Fe_2B coatings. The boron concentration profiles at the Fe_2B coating is depicted in figure 1. $C_{up}^{Fe_2B}$ and $C_{low}^{Fe_2B}$, are the upper and lower limits of boron concentration in the Fe_2B coatings. The boride layers have narrow boron concentration ranges of about 35×10^{-4} wt.%B [5]. Since only Fe phase is present before boriding, the initial condition of the diffusion problem is given by the following equation [7]:

$$C_{Fe_2B}\left[x(t > 0) = 0\right] = 0 \tag{1}$$

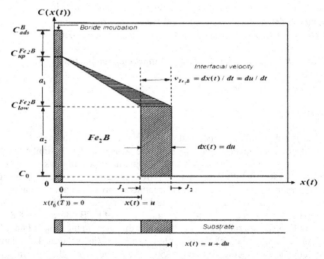

Figure 1. Boron concentration profiles along the interfaces Fe_2B/substrate.

Also, the boundary conditions (Figure 1) of the diffusion problem are given by the following equations [8-9]:

$$C_{Fe_2B}\left[x(t=t)=0\right]=C_{low}^{Fe_2B} \text{ for } C_{ads}^B < 8.83 wt.\%B \tag{2}$$

$$C_{Fe_2B}\left[x(t=t_o)=0\right]=C_{up}^{Fe_2B} \tag{3}$$

$$C_{Fe_2B}\left[x(t=t)=u\right]=C_{low}^{Fe_2B} \tag{4}$$

$$C_{Fe_2B}\left[x(t=t)=u\right]=C_0 \tag{5}$$

Where C_{ads}^B is the effective adsorbed boron concentration, while $8.83 wt.\%B$ is the boron composition at the Fe$_2$B coating. The phenomenon of mass transfer at the growth interface (Fe$_2$B/substrate) is described by the mass balance equation [9]:

$$\Delta_{Fe_2B} = a_2 du + \frac{1}{2}a_1 du = J_1 dt - J_2 dt \tag{6}$$

Where from Fig. 1, $a_1 = C_{up}^{Fe_2B} - C_{low}^{Fe_2B}$ defines the homogeneity range of the phase designed Fe$_2$B, $a_2 = C_{low}^{Fe_2B} - C_0$ is the miscibility gap and C_0 is the terminal solubility of the interstitial solute. The concentrations $C_{up}^{Fe_2B}$, $C_{low}^{Fe_2B}$ and C_0 are expressed in wt.%B. The boron concentration profiles at the Fe$_2$B coating is given by the following equation:

$$C_{Fe_2B}\left[x(t)\right]=C_{up}^{Fe_2B} + \frac{C_{low}^{Fe_2B} - C_{up}^{Fe_2B}}{u}x(t) \tag{7}$$

The fluxes J_1 and J_2 are then given by Fick's first law, $J = -D\{dC[x(t)]/dx(t)\}$ as follows:

$$J_1 = D_{Fe_2B}a_1 / u \tag{8}$$

Where D_{Fe_2B} is the boron diffusion coefficient in the Fe$_2$B layer, Assuming that the substrate becomes saturated within a short period of time, and boron solubility is extremely low in $\gamma - Fe$ (0.003 wt.%B), $J_2 \cong 0$.

Substituting Eq. 8 into Eq. 6, the mass balance equation at the growth interface is then given by the following equation.

$$(a_2 + a_1 / 2)\frac{du}{dt} = D_{Fe_2B}\frac{a_1}{u} \tag{9}$$

The Fe$_2$B coating growing obeys the power law, of the form:

$$u(t) = k\left\{t^{1/2} - \left[t_0(T)\right]^{1/2}\right\} \tag{10}$$

Where u indicating average thickness of the Fe$_2$B coating, t corresponding to treatment time, $t_0(T)$ is the incubation period required for the formation Fe$_2$B coating as a function of the temperature, and k is the parabolic growth constant. Therefore, using the Eq. 9 the boron diffusion coefficient at the Fe$_2$B phase (D_{Fe_2B}) is determined as:

$$\frac{1}{P}\int_0^u u\,du = D_{Fe_2B} \int_{t_0(T)}^t dt \tag{11}$$

with $P = \left(2a_1/(2a_2+a_1)\right)$, based on the aforementioned assumptions, the boron diffusion coefficients in the Fe$_2$B coating, (D_{Fe_2B}) may be expressed using equation 11 as follows:

$$D_{Fe_2B} = \frac{k^2}{2P}\left\{ \frac{\left(1-[t_0(T)/t]^{1/2}\right)^2}{1-[t_0(T)/t]} \right\} \tag{12}$$

This permits the determination of D_{Fe_2B} in terms of known parameters P , a_1 and a_2 are obtained from the Fe–B phase diagram [5-10-11], while a_1 =0.17 and a_2 =7.23, (values in wt.% B). The values of k are given by the slopes of the least-squares fitted lines from the plots of u versus $t^{1/2}$.

EXPERIMENTAL

The samples ductile iron ASTM A-536 class 80-55-06 were used as substrate, with a nominal composition of 3.60-3.90%C, 2.30-2.80 %Si, 0.10-0.30%Mn, 0.10 %P, 0.015 %S, 0.043%Cr. The samples cubes of 10x10x50 mm. The process was carried out at temperatures of 1173, 1223 and 1273 K with three exposure times of 6, 7, and 8 h. Cross section of the sample were prepared by standard metallographic, and the depth of the surface layers was observed using JEOL JSM 6360 LV scanning electron microscope (SEM). In order to minimize the roughness effect at the growth interfaces fifty measurements were made on different sections of the ductile iron boriding samples to estimate the average length of the Fe$_2$B coating. In addition, X-ray diffraction (XRD) was carried out on the boride sample obtained at a temperature of 1273 K with 8 h of exposure.

RESULTS AND DISCUSSION

Figure 2 shows SEM cross-sectional examination of boronized ASTM A-536 ductile irons, Fe$_2$B coating formed on the surfaces. Furthermore, cross section also revealed the presence of characteristic saw-tooth morphology, in which the Fe$_2$B coating includes globular graphite in their matrix.

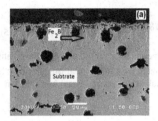 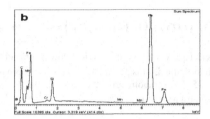

Figure 2. (a) SEM image of ferrite-pearlite of microstructure with morphology Fe$_2$B layers formed at the surface of ductile iron ASTM A-536 boriding for 6 h exposure at 1223 K and (b) EDS spectrum of boride sample.

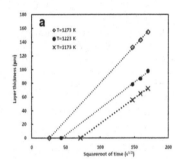 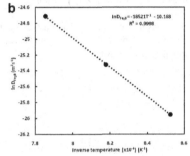

Figure 3. (a) Growth kinetics of layers Fe$_2$B at the surface of ductile iron ASTM A-536 (b) Behavior of the boron diffusion coefficient in Fe$_2$B phase as a function of the boriding temperature

The Growth boride is shows figure 3 (a). The boron diffusion in Fe$_2$B coating at each boriding temperature was evaluated using equation 12. The boron diffusion coefficients are calculated using the Arrhenius relationship as shown in figure 3 (b) and are thus:

$$D_{Fe_2B} = 3.84x10^{-05} \exp\left(-\frac{155.7\,kJ\,mol^{-1}}{RT}\right) \ (m^2s^{-1}) \tag{13}$$

Where R is the universal gas constant 8.314 J/mol.K, the T represent the absolute temperature and Q_{Fe_2B} the activation energy values obtained from the slopes of the straight lines of figure 3 (b) represent the energy necessary to stimulate the diffusion of boron in the [001] direction along the Fe$_2$B coating. To stimulate automation of the boriding process, it is necessary to correlate boron surface content and boride incubation time with the parabolic constant k. Hence from Eq. 12, the square parabolic growth constant may be estimated as:

$$k^2 = 2P\left\{1-\left(t_o(T)/t\right)\Big/\left(1-\left(t_o(T)/t\right)^{1/2}\right)^2\right\}D_{Fe_2B} \tag{14}$$

And
$$\varepsilon(T) = \left\{ 1 - \left(t_o(T)/t \right) \middle/ \left(1 - \left(t_o(T)/t \right)^{1/2} \right)^2 \right\} = -0.0131T + 18.045 \tag{15}$$

The $\varepsilon(T)$ parameter has no physical dimension and depends only on the process temperature, as shown in figure 5. This parameter can also be approximated by a linear relationship given by Equation 15. The parabolic growth constant at the (Fe_2B/substrate) interface can be rewritten, expression in $\left(\mu m s^{-1/2} \right)$ as:

$$k^2 = 2 P \varepsilon(T) D_{Fe_2B} \quad \mu m s^{-1} \tag{16}$$

Where $\varepsilon(T) = -0.0131T + 18.045$ and $P = 0.0191 wt.\% B$

$$k = \left((-5.004 x 10^{-04} T + 0.6893) D_{Fe_2B} \right)^{1/2} \quad \mu m s^{-1} \tag{17}$$

The growth constants can be predicted using equation 17, with the influence of boron surface content P and the $\varepsilon(T)$ parameter. The results agree with experimental parabolic growth constants obtained in this work.

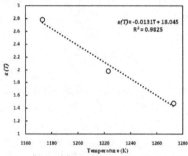

Figure 4. Variation of $\varepsilon(T)$ parameter with the boriding temperature.

Using the Eq. 16 and substituting in Eq. 10, the layer thickness of the Fe_2B coating (u) is given as:
$$u = \left(2 P \varepsilon(T) D_{Fe_2B} \right)^{1/2} \left\{ t^{1/2} - \left[t_0^{Fe_2B}(T) \right]^{1/2} \right\} \tag{18}$$

The time incubation $t_0^{Fe_2B}(T)$ ratio varies linearly with the process temperature. It can be checked that the boride incubation time is reduced when the temperature increases. At a given temperature a correlation can be established between the boride incubation time and treatment time according to

$$\eta(T) = 1 - \left[t_0(T) / t \right]^{1/2} = 1.7 x 10^{-3} T - 1.1808 \tag{19}$$

Where $\eta(T)$ is a parameter which has no physical dimension, $\eta(T)$ versus the temperature follow a linear relationship as show in figure 5. Combining Eqs. 15 and 19, the $\varepsilon(T)$ parameter can be determined, as described by Ortiz-Dominguez et al. [12]:

$$\varepsilon(T) = \frac{2 - \eta(T)}{\eta(T)} \tag{20}$$

So, the predicted Fe$_2$B coatings presented in equation 18 given by:

$$u = \left(2P[2 - \eta(T)] \eta(T) D_{Fe_2B} t \right)^{1/2} \tag{21}$$

The results obtained in Eq. 21 are compared with experimental values obtained in the present study as shown in Table I. Equation 21 can be used as a simple tool to predict the Fe$_2$B boride layers thicknesses of ductile iron ASTM A-536 according to their practical applications and can also serve to select the optimized parameters for boriding process. The validate the diffusion model, sample of ductile cast iron were boride at the temperature of 1273 and 1123 K with 10h of treatment time. The estimated values obtained by Eq. (21), show good agreement with the experimental results (see table I).

Table I. Experimental and predicted values of the boride layer thickness in the Fe$_2$B phase for different exposure times at temperatures.

Time (h)	Temperature 1173 (K)		Temperature 1273 (K)	
	Experimental (μm)	Predicted from Eq. 21 (μm)	Experimental (μm)	Predicted From Eq.21 (μm)
10	72.04	75.385	105.01	104.610

CONCLUSIONS

In this work the growth kinetics of boride layers formed at the surface of ductile irons ASTM-A536 have been estimated. Fe$_2$B coatings formed swan morphology and was developed without the formation of discontinuous graphite phases at the interface Fe$_2$B/substrate. The parabolic growth constant k and thickness of Fe$_2$B coating u were described through $\varepsilon(T)$ and $\eta(T)$ parameters, depends on the boriding temperature and boride incubation time $t_0(T)$, respectively. Additionally, the velocity of the Fe$_2$B/substrate interface and the weight gain at the sample boride are evaluate using parameters $\varepsilon(T)$ and $\eta(T)$. The results indicate that under the applied boride condition, good agreement was obtained between the experimental and predicted boride hard coating. Consequently, automation and optimization of the pack-boriding process in ductile irons studied in this work can be achieved.

ACKNOWLEDGEMENTS

Authors gratefully acknowledge the support given by Consejo Nacional de Ciencia y Tecnologia, CONACyT through the Doctoral Program of Instituto Politecnico Nacional and Promep.

REFERENCES

1. A. K. Sinha, "Boronizing", ASM Handbook, OH, USA, Journal Heat Treat. 4, (1991), p. 437-447.

2. A. Graf Von Matuschka, "Boronizing", First ed., Carl Hanser Verlag, Germany, (1980), p. 31-40.

3. J. R. Davis, "Surface hardening of steels: understanding the basics", first ed.ASM International, USA, (2002) p. 213-215.

4. Mourad Keddam et. al. Solid Comp. of Trans. Elem. 170, 185-189 (2011).

5. T. B. Massalski, Binary alloy phase diagrams. I and II, 482-1273 (1986).

6. T. Van Rompaey, K. C. Hari Kumar, P. Wollants, J. Alloys Compd. 334, 173–181(2002).

7. L.G. Yu, X.J. Chen, K.A. Khor, G. Sundararajan , Acta Mater, 53, 2361- 2368 (2005).

8. M. A.J. Somers, E. J. Mittemeijer, Metall. Trans. A, 26, 57-74 (1995).

9. Mourad Keddam, Defect and Diffusion Forum, 269, 297-301 (2010).

10. B. Hallemans, P. Wouants, and J.R. Roos, Z. Metallkd, 85, 10 676-682 (1994).

11. H. Okamoto, J. Phase Equil. Diffusion, 25, 297-298 (2004).

12. M. Ortiz-Domínguez, E. Hernández-Sánchez, J. Martínez-Trinidad, M. Keddam, I. Campos, Kovove Mater. 48, 285–290 (2010).

Mater. Res. Soc. Symp. Proc. Vol. 1481 © 2012 Materials Research Society
DOI: 10.1557/opl.2012.1639

Luminescence in a Ba(Ti,Zr)O$_3$ Films Deposited by Ultrasonic Spray Pyrolysis Method

D. Y. Medina[1*], R. T. Hernandez[1], I. Hernandez[1], S. Orozco[2]

[1]División de Ciencias Básicas e Ingeniería, UAM-A, Av. Sn Pablo 180, 02200 México D.F., México.

[2]Depto. de Física, Facultad de Ciencias UNAM, Av. Universidad 3000, Col. Copilco el Bajo, México D.F., México.

*E-mail: dyolotzin@correo.azc.uam.mx

ABSTRACT

The Ba(Ti,Zr)O$_3$ powders was synthesized for many methods because its properties as a piezoelectric, dielectric and ferroelectric material that insert it as an functional material, but there is little information in the studied of its optical properties. In this work Ba(Ti,Zr)O$_3$ films were produced by ultrasonic spray pyrolysis method for optical applications. The precursors used were an barium acetyl-acetonate , titanium acetyl-acetonate , and zirconium acetyl-acetonate powders dissolved in a N-N, dimethylformamide solution. Optical and morphological properties of the films shown a non crystalline structure and its emission spectra shown a broad and intense luminescence at 468nm which correspond to visible emission in the green region.

Keywords: amorphous, luminescence, film, photoemision, spray pyrolysis.

INTRODUCTION

In the recent years several studies have reported room temperature visible photoluminescence (PL) in structurally disordered materials in the form of non crystalline films [1]. In other hand there are an increasing use for light-emitting devices for displays and communication, light-emitting diodes (LEDs) and many optical and electronic applications that generate an increasing demand of new photoluminiscent materials at low cost. The Ba(Ti,Zr)O$_3$ is a perovskite material in crystalline structure and it has been synthesized for many methods. It has properties as a piezoelectric, dielectric and ferroelectric material. As a thin films are expected to be used in microelectronic devices [2-5]. That insert it as a functional material, but there is little information in the studied of its optical properties [6, 7].

Various methods for synthesis of luminescent powders and films have been employed [8-11]. The ultrasonic spray pyrolysis method is a well established process for depositing films. The main advantages of this technique are its low cost, a high deposition rate, the possibility to coat large areas, its ease of operation and the quality of the coatings obtained which can be easily scaled to industrial level [12-17]. In his work are presented the synthesis and characterizations of Ba(Ti,Zr)O$_3$ films obtained with ultrasonic spray pyrolysis technique.

EXPERIMENTAL PROCEDURE

The films deposition procedure is similar to previously reported [1].The films were grown over glass slides of Corning 7059 type (commercial grade) like substrates. Were used as source materials acetylacetonates (Acac of Sigma Aldrich) of Ba, Ti and Zr dissolved in N,N,dimetylformamide (Dmf of Sigma Aldricht) to form the spraying solution with different concentrations. The substrates were placed on a hot plate at constant temperature (T_S) between 450 to 600°C and a spray nozzle was located approximately 6mm above the substrate surface, all enclosed in an exhaust hood. After experimental analysis the best deposition conditions which give the homogeneous films were: The spraying solution consists in a 0.02M Ba-Acac, 0.015M Ti-Acac and 0.01M Zr-Acac in a 100 ml of Dmf. The carrier gas (filtered air) was 3200ml/min. The deposit time was 15 min and the ultrasonic nebulizer (commercial type) at 0.9 Mhz vibration. The films were deposited at Ts of 600 °C.

Measurements of photoluminescence (PL) were performed at room temperature and atmospheric pressure. The emission spectra was recorded using a spectrograph: Fluoro Max-P Jobin Yvon (slit with: 3nm), and an excitation wavelength of 270nm. The structure of the films was analyzed by X-ray diffraction (XRD) pattern to razing angle at room temperature: using a Siemens 5000. The surface morphology of the films was analyzed by scanning electron microscopy (SEM): JEOL JSM 6400. The film thicknesses were measured with a perfilometer: *KLA* Tencor.

RESULTS AND DISCUSSION

Figure 1 show the XRD spectrum pattern to razing angle obtained from a film grown at 600°C, which correspond to a practically amorphous structure. With a little changes, similar XRD spectrums are obtained in all temperature range of deposition. That result is new and has not been reported and is associated to the grown technique, particularly to the temperature of deposition. A film average thickness of 8μm was obtained from the perfilometer test. The SEM image in Figure 2 also is of the sample prepared at 600°C showing that the surface film are rough and uniform across the covered area.

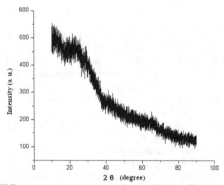

Figure 1. The XRD spectrum pattern to razing angle for a film grown at 600°C

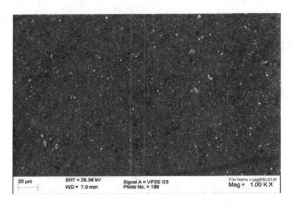

Figure 2. Micrograph by SEM of a film deposited at 600°C.

Figure 3 shows the film grow at 600°C the most important emission transition of the Ba(Ti,Zr)O₃ at 468nm which corresponded to the green color in the visible region and secondary transitions was founded at 452, 480, 492nm. Results for the amorphous films obtained. However, the results are according with founded by X. G. Tang, et al and L.S. Cavalcante *et* al. [7, 18]. Who found that the emission is related to the degree of order–disorder in the Ba(Ti,Zr)O₃ thin film lattice. When the disorder of the lattice was increased the peaks were displaced to regions with lower wavelength. In this case is not has a crystalline lattice has an amorphous lattice.

A secondary transition was found at 825nm which correspond to the last region of the visible in the red color and maybe a transition in de near infrared region which can be observed in Figure 4. This transition made that the Ba(Ti,Zr)O₃ films can be considered for infrared red LED's production.

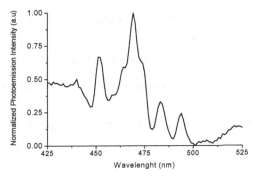

Figure 3. Photoemission spectrum in the range of 425-525nm of a film deposited at 600°C.

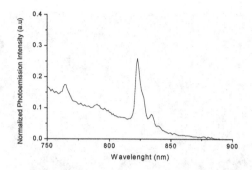

Figure 4. Photoemission spectrum in the range of 750-900nm of a film deposited at 600°C.

CONCLUSIONS

With spray pyrolysis method can been produced a Ba(Ti,Zr)O$_3$ amorphous films. The Ba(Ti,Zr)O$_3$ is an excellent luminescent material. The films obtained can be effectively excited by ultraviolet (270 nm) light, and emit in the region of green light, other secondary emission was found at 825nm in near infrared region, which made that the films can be used in Light Emitting Diode (LED) production.

REFERENCES

1. D.Y. Medina, R.T. Hernandez, I. Hernandez, S. Orozco *Journal of Non Crys-Solids,* **357**, 3740–3743 (2011).

2. P. Zurlini, A. Lorenzi, I. Alfieri, G. Gnappi, A. Montenero, N. Senin, R. Groppetti, and P. Fabbri, *Thin Solid Films,* **517**, 5881 (2009).

3. N. Binhayeeniyi, P. Sukvisut, C. Thanachayanont, and S. Muensit, *Mater Lett,* **64**, 305 (2010).

4. K. Mimura, T. Naka, T. Shimura, W. Sakamoto, and T. Yogo, *Thin Solid Films,* **516**, 8408 (2008).

5. R. Vivekanandan, S. Philip, and T. R. N. Kutty, *Mater. Res. Bull.* **22**, 99 (1987).

6. S. K. Rout, L. S. Cavalcante, J. C. Sczancoski, T. Badapanda, S. Panigrahi, M. Siu Li, and E. Longo, Physica B: *Condensed Matter,* **404**, 3341 (2009).

7. X. G. Tang, H. L. W. Chan, and A. L. Ding, *Thin Solid Films,* **460**, 227 (2004).

8. G. Strasser, E. Bertel, and F. P. Netzer, *Journal of Catalysis*, **79**, 420 (1983).

9. A. Ohkubo, A. Ohtomo, J. Nishimura, T. Makino, Y. Segawa, and M. Kawasaki, *Appl. Surf. Sci.* **252**, 2488 (2006).

10. A. Kirianov and A. Yamaguchi, *Ceram. Int.*, **26**, 757 (2000).

11. J. Zhao, X. Li, J. Bian, W. Yu, and C. Zhang, *Thin Solid Films*, **515**, 1763 (2006).

12. T. Kodas T. and M. Hampden-Smithnd J., *Aerosol Processing of Materials*, (Wiley-VCH, New York, 1999).

13. R. Martínez-Martínez, E. Álvarez, A. Speghini, C. Falcony, and U. Caldiño, *Thin Solid Films*, **518**, 5724 (2010).

14. R. Martínez-Martínez, M. García-Hipólito, L. Huerta, J. Rickards, U. Caldiño, and C. Falcony, *Thin Solid Films*, **515**, 607 (2006).

15. A. Ortiz, M. Garcia, and C. Falcony, *Mater. Chem. Phys.*, **24**, 383 (1990).

16. Y. Iwako, Y. Akimoto, M. Omiya, T. Ueda, and T. Yokomori, *J. Lumin*, **130**, 1470 (2010).

17. K. Y. Jung, D. Y. Lee, Y. C. Kang, and H. D. Park, *J. Lumin.* **105**, 127 (2003).

18. L. S. Cavalcante, M. F. C. Gurgel, E. C. Paris, A. Z. Simões, M. R. Joya, J. A. Varela, P. S. Pizani, and E. Longo, *Acta Materialia*, **55**, 6416 (2007).

Mater. Res. Soc. Symp. Proc. Vol. 1481 © 2012 Materials Research Society
DOI: 10.1557/opl.2012.1640

Characterization on Fracture Surfaces of 304 Stainless Steels Joined by Brazing Using Silicon Nanoparticles

L. Santiago-Bautista, H. M. Hernández-García, R. Muñoz-Arroyo, M. Garza-Castañón, F. García-Vázquez, J. Acevedo-Dávila

Corporación Mexicana de Investigación en Materiales S.A. de C.V., Ciencia y Tecnología No. 790 Fracc. Saltillo Coah., C.P. 25290, México.

E-mail: lsantiago@comimsa.com.

ABSTRACT

Silicon nanoparticles of 100 nm obtained by high-energy ball milling were characterized by X-ray diffraction (XRD) and transmission electronic microscopy (TEM). Results show dark areas due to a staking of defects. On the other hand, brighter areas exhibit a combination of small crystalline and amorphous zones. To fulfill and cover the micro-cracking and micro-pores generated during the welding process of 304 stainless steels joined by brazing, these nanoparticles were deposited directly in the fracture. The amorphous silicon drove the Transient Liquid Phase (TLP) at 1000°C for 20 min. This amorphous silicon decreases the energies of reaction between the substrate and melting filler. TLP increases the wettability and capillary forces between micro-cracking and micro-pores; due to that, the eutectic phase contained by the melting filler forms a liquid. Moreover, the weld beads were characterized by Scanning Electron Microscopy (SEM) to analyze the effect of silicon nanoparticles on the weld beads. These results showed that the interaction of the Si nanoparticles with metallic filler in the melting zone decreases the size and change the morphology of the present phases as well as the zone of isothermic growth.

Keywords: welding, steel, nanostructure, brazing, fracture.

INTRODUCTION

Stainless steels are essential for medical, chemical, food, and biotechnological applications because of their excellent properties such as corrosion resistance, weldability and hardness at room temperature. However, under certain conditions, are susceptible to corrosion, which reduces their mechanical properties giving place to fracture initiation. These adverse effects are mainly present near the areas joined by welding. Micro cracking without soldering is mainly because the flux used during the brazing process does not adequately penetrates cracks or spaces given their complex morphologies. Therefore, binding has to be used after the formation of a transient liquid phase (TLP). This is a new process that welds base material with an intermediate layer. The intermediate layer is melted and the inserted element diffuses into the substrate material, causing isothermal solidification [1-3]. If the isothermal solidification is not completed, the liquid in the intermediate layer is solidified through eutectic phases during cooling, causing a decrease of the mechanical properties of the weld bead. Xiaowei Wu et al [4] reported that TLP bonding process is carried out in four steps: 1) melting of the intermediate layer, 2) dissolution of base metals (in thermodynamic equilibrium solid-liquid interface) 3) isothermal solidification and 4) homogenizing the isothermal zone of growth. Furthermore, alloys containing substantial contents of chromium are highly susceptible to show cracking in the heat affected zone (HAZ) near the weld bead [5]. Therefore, the filler metals containing silicon and boron decrease the melting point and reduce the eutectic structures [5, 6]. Moreover, the presence of the TLP, rich in silicon and boron, prevents damage to the weld bead [7, 8]. The objective of this work is the to study the formation of transient liquid phases on fractures of 304 stainless steel, to promoting the following: a) forming a liquid capable of increasing the wettability between microcracks and internal micropores, b) improving capillary forces between the filler metals and the transient liquid phase and c) modifying the phase formation.

EXPERIMENTAL

Obtaining silicon nanoparticles and fractures of 304 stainless steels joints

100 nm sized silicon nanoparticles were obtained from Si powder of 35 μm Aldrich brand using high energy milling at 350 rpm for 8 h. The nanoparticles were then characterized by XRD and HRTEM FEI Titan. Fractures of 304 stainless steels were obtained for 10x60 mm rods by bending using a mechanical testing machine brand. The fracture surfaces were inspected by SEM.

Characterization of the filler metal and impregnation of Si nanoparticles on fractures of stainless steels

Before carrying out the brazing process of the stainless steel rods, the BNi-9 filler metal was characterized by XRD and SEM in order to establish the chemical composition and size of the agglomerate. Subsequently, the impregnation of Si nanoparticles was carried out on the fracture surface of the base metals. A Mixture of 0.5 g of Si nanoparticles in 200 ml of ethanol was made. Nanoparticles were then dispersed for 1 h in an ultrasound machine Brason 2510 brand. Fractures of steel were placed in ethanol with the dispersed nanoparticles for 30 min in ultrasound machine. These times were proposed to ensure that the nanoparticles were impregnated inside the micropores and microcracking.

Brazing joining of steels with and without nanoparticles of Si

The study of the effect of the Si nanoparticles in the brazing process of the stainless steel was carried out for two cases: first case) determining the degree of reactivity of the Si nanoparticles at 1000°C and 1200°C for different times on the fracture surface and microcracking without using BNi-9 filler metal and second case) Brazing joining of fractures of stainless steels without and with nanoparticles using the BNi-9 filler metal. In both cases, the brazing process was conducted in a sealed tube furnace at brazing temperatures of 1200°C for 60 min under a 100 ml/min flow of Ar gas. Using 10°C/min heating and cooling rates. The surfaces of fractures with nanoparticles and without BNi-9 filler metal as well as the weld bead in the samples with and without nanoparticles and BNi-9 were characterized by SEM.

RESULTS AND DISCUSSION

Getting silicon nanoparticles and fracture of 304 stainless steels

Figure 1 shows the XRD pattern of Si nanoparticles. Intense peaks of Si are observed and an amorphous-material-like behavior. Probably due to the nanometric size of Si, this was evidenced by HRTEM (Figure 2).

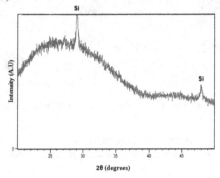

Figure 1. Pattern of XRD of Si nanoparticles obtained by high mechanical milling for 8 h.

Figure 2 shows 50-100 nm sized Si nanoparticles. It can be seen at higher amplifications (Figure 3) two types of materials: a) Amorphous material and b) dark areas showing high density of dislocations mainly due to the mechanical action of the milling process. The condition of amorphous phase of the nanoparticles allows chemical reactions that require less energy for the formation of transient liquid phases inside the microcracks. For the case of high density of dislocations, these are high energy zones for nucleation and growth phases.

Figure 2. HRTEM image of Si nanoparticles.

121

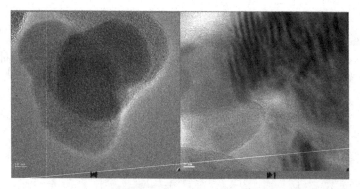

Figure 3 HRTEM images of Si. a) Amorphous Nanosilicon b) Nanoparticles with high density of dislocations.

Figure 4 shows the fracture surface and tear of material inside a micropore of a 304 stainless steel. Both photomicrographs show ductile fracture and crack propagation due to coalescence and growth of pores.

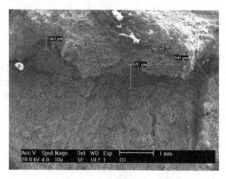

Figure 4. Secondary electron images of the ductile fracture surface of a stainless steel 304.

Characterization of the BNi-9 filler metal and impregnation of Si nanoparticles on fractures

Figure 5 shows an XRD pattern and SEM of the BNi-9 filler metal. Basically, the chemical nature is predominantly composed of $FeNi_3$, (Fe, Ni) $_{23}C_6$, and Ni_3B. This last compound is an eutectic of low melting point [8]. Reports in the literature [2] establish a residence time of fluid in the isothermal growth zone of less than 1 s. Moreover, the agglomerates of spherical morphology of the BNi-9 filler metal show an average size of 5-50 microns. This represents a technical problem for joining the components with less than 1 micron microcracks. Because the penetration of the filler metal is not effective enough to fulfill the space. This reduces the efficiency of the repairing process of the component.

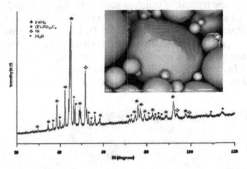

Figure 5. XRD Pattern and SEM of the BNi-9 filler metal.

As mentioned, this work focuses on the formation of transient liquid phases of Si nanoparticles using impregnation technique for 30 min on fracture surfaces and interior of the microcracking to promote the interaction with the compounds of the filler metal, especially with the low melting point eutectic (NiB₃), Thus activating the capillary forces in the interior of the microcracking and micropores. The growth kinetics of the phases is a function of the thermal cycle of brazing and subsequent heat treatments.

<u>Brazing joining of steels with and without Si nanoparticles</u>

First case: determining the degree of reactivity of the nanoparticles at 1000°C and 1200 °C for different times on the fracture surface and microcracking without using BNi-9 filler metal. Figure 6 shows the formation of the transient liquid phase rich in Si at 1000°C for 20 min. It can be seen the occupation of the liquid phase inside the micropores without the use of filler metal due to the amorphous nanoparticles obtained by milling.

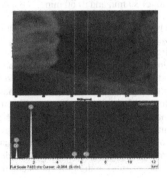

Figure 6. SEM image and ESD of the transient liquid phase formed at 1000°C for 20 min inside a micropore.

Figure 7 (a), (b), (c) and (d) shows the degree of reactivity of nanoparticles of Si at 1200 ° C versus time without adding the BNi-9 filler metal. In addition, it can be observed the evolution of growth and morphology on the fracture of stainless steel. In the Figure 7 (a) shows the formation of a SiO_2 rich core surrounded of solid particles at 10 min. At 20 min can be seen the crystal growing on solidified transient liquid phases (Fig. 7 (b)). By contrast, at 30 min the liquid phases are saturated of chemical species that give rise to the formation of large semi-spherical particles (Fig. 7 (c)). At 60 min can be seen the growth and breaks the surface of the semi-spheres. Only were observed extended branches as of a ring that to lock up a core rich of SiO (Fig. 7 (d)). These figures were taken at different amplifications due to that are clear to explain the phenomena.

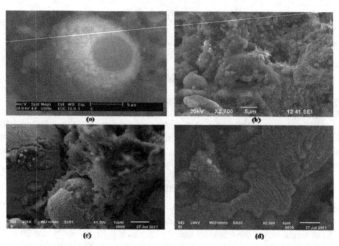

Figure 7. SEM image of fracture surfaces of stainless steels previously coated with Si nanoparticles. a) 10 min, b) 20 min, c) 30 min and d) 60 min.

Second case: Stainlees steels joined with and without Si nanoparticles using the BNi-9 filler metal. Figure 8 (a) and (b) shows the zones of the stainless steel joined by brazing process at 1200°C for 60 min without and with nanoparticles of Si, respectively. Both cooled inside the pipe furnace (10°C / min). As can be seen in the samples without nanoparticles (Fig. 8 (a)), the growth of the isothermic zone and the intermetallics in the melting zone. Additionally, a marked precipitation zone. In contrast, in the figure 8 (b) has a narrow isothermic growth zone stages as well as a change in the morphology and sizes of the intermetallics. Because of the effect of Si nanoparticles previously impregnated on the fracture surface of 304 stainless steel. These facts inspected in the samples with Si nanoparticles suggest following: a) in spite of the bonding gap is too wide, the diffusive effect of Si for 60 min changes the morphology and size of the intermetallics and b) the isothermic zone is reduced due to the decrease of the growth rate between the chemical species (Boron, mainly) in the filler metal and the Si nanoparticles.

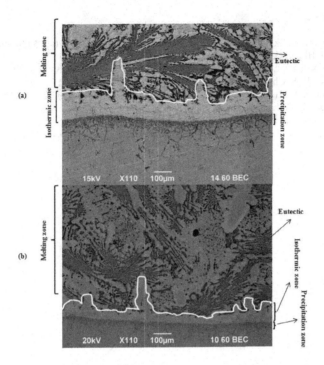

Figure 8. Backscattered electron images of the weld bead in the stainless steel joined by brazing at 1200°C for 60 min. a) without nanoparticles and b) with nanoparticles.

CONCLUSIONS

The time of impregnation time by 30 min of silicon nanoparticles is sufficient to fill the inside of the micropores and intricate spaces of microcrackings where the filler metal cannot penetrate. At 1000°C for 20 min transient liquid phases are formed with the addition of Si nanoparticles. Obviously, the identification of other trace elements is due to chemical reaction with the metal base. Moreover, in the samples without filler metal and Si nanoparticle at 1200°C for 60 min increase its size and acquire a surface appearance of a semi-sintered spherical morphology. This follows a high reactivity of nanoparticles by thermal effect. Finally, in fractures joined by brazing at 1200°C for 60 min, shown in the melting zone of fractures without nanoparticles an acicular microstructure. In contrast, fractures joined by Brazing and Si nanoparticles modified the eutectic. The use of the nanotechnology can be a way to change the size and morphology of the phases taking in account a gap of 1 mm. and to obtain better mechanical properties on the weld bead.

REFERENCES

1. M. Pouranvari, A. Ekrami and A. H. Kokabi, *J. Alloys Compd.* **469**, 270-275 (2009).
2. G. O. Cook III and C. D. Sorensen, *J. Mater. Sci.* **46**, 5305-5323 (2011).
3. N. R. Philips, C. G. Levi and A. G. Evans, *Metall. Mater. Trans. A* **39A**, 142-149 (2008).
4. W. Xiaowei, R. S. Chandel, L. Hang, *J. Mater. Sci.*, **36**, 539-546 (2001).
5. R. K. Sidhu, N. L. Richards and M. C. Chaturvedi, *Mater. Sci. Technol.* **5**, 529-539 (2008).
6. M. J. Nakahashi, *J. Instr. Metall.* **4**, 285-290 (1987).
7. J. S. Jang and H. P. Shih, *J. Mater. Sci. Lett.* **22**, 79-82 (2003).
8. N. R. Philips, C. G. Levi, and A. G. Evans, *Metall. Mater. Trans. A* **39A**, 143-149 (2008).

Mater. Res. Soc. Symp. Proc. Vol. 1481 © 2012 Materials Research Society
DOI: 10.1557/opl.2012.1641

Synthesis of ZnO at Different Atomic Proportion Produced by Chemical Precipitation

A. Medina[1,4,*], L. Béjar[2], G. Herrera-Pérez[3]

[1]UMSNH, Instituto de Investigaciones Metalúrgicas, Edificio U Ciudad Universitaria, C.P. 58040, Morelia, Michoacán, México

[2]UMSNH, Facultad de Ingeniería Mecánica, Edificio W Ciudad Universitaria, C.P. 58040, Morelia, Michoacán, México

[3]Departamento de Ingeniería en Materiales, Instituto Tecnológico Superior de Irapuato (ITESI) Carretera Irapuato-Silao Km. 12.5, El Copal, Irapuato, Guanajuato. C.P. 36821, México

[4]SEP-DGEST-IT de Tlalnepantla, Av. Tecnológico s/n, Col. la Comunidad, Tlalnepantla de Baz, Edo México, 54070, México.

*E-mail: ariosto@umich.mx

ABSTRACT

Zinc oxide (ZnO) nanoparticles were produced using chemical precipitation synthesis with a molar ratio of 1:1, 1:2 and 1:3. The structure, chemical composition and morphology were investigated by X-ray diffraction (XRD), energy dispersive spectroscopy (EDS), scanning electron microscopy (SEM), transmission electron microscopy (TEM) and high resolution transmission electron microscopy (HRTEM). XRD and EDS demonstrated that the all particles formed at different atomic proportion were of wurtzite crystal structure with the same chemical composition. SEM and TEM showed the formation of hexagonal particles with a molar ratio of 1:1 while the samples synthesized with a molar ratio 1:2 and 1:3 showed a circular shape. HRTEM and Fast Fourier Transform (FFT) demonstrated that the all particles were formed with a preferable [0001] growth direction.

Keywords: Chemical synthesis, scanning electron microscopy (SEM), nanostructure, crystal, grain size.

INTRODUCTION

In the past few years, one dimensional structure has generated the most excited in advanced-materials research because of their fascinating properties from both scientific and engineering-oriented standpoints. ZnO is a piezoelectric and semiconducting material having wide band gap of 3.37 eV and potential applications used in nanoscale dye-sensitized solar cells [1,2] optoelectronics [3,4] gas sensors [5,6] and electronics [7,8]. ZnO is a versatile material with probably the broadest spectrum of applications ranging from optoelectronics to catalysis to sensors to cosmetic to nanomedicine [9,10]. Controllable synthesis of nanomaterials is very important to nanoscale science, which denotes fabrication on arrays with desirable morphology [11]. Arrays of ZnO exhibits diverse morphologies such as belts, tubes, rods [12], however, it is difficult up date to control the morphology of the structures through one synthesis process. Therefore, extensive research on the quasi-one-dimensional (1D) ZnO nanostructures has been conducted recently. Diverse fabrication techniques, such a thermal evaporation, molecular beam epitaxy, vapor phase transport, chemical vapor deposition (CVD), metal organic, aqueous solution deposition, and electrochemical deposition, have been attempted to grow the materials [11]. The significances of the controllability manifest in both the chemistry of small-size material synthesis and the realization of their applications. In this communication we present the characterization high quality arrays of ZnO obtained directly by chemical precipitation. For this synthesis we consider three cases based on the stoichiometric ratio between $Zn(H_3C_2O_2)_2$ and NaOH precipitating agent. (Limit reactant NaOH (1:1), stoichiometric (1:2) and the NaOH excess reagent (1:3)).

EXPERIMENTAL PROCEDURE

In the reaction a-$Zn(H_3C_2O_2)_2 \bullet 2H_2O_{(aq)} + b$-$NaOH_{(aq)} \rightarrow Zn(OH)_{2(s)} \downarrow + 2Na(H_3C_2O_2)_{(aq)}$, a and b are the stoichiometric coefficients, the relation for each case considered in this work will be indicated as $a:b$. In the first step 40 g of $Zn(H_3C_2O_2)_2 \bullet 2H_2O$ was dissolved in 200 mL of distilled water, and stirred at 475 rpm a constant temperature of 40 °C. A second solution was prepared by dissolving 7.3 g of NaOH in 200 mL of distilled water and added to the Zn-solution with a speed of 4 mL/min and stirred for 30 min. The precipitates are centrifuged and washed with distilled water at room temperature to remove ionic species of Na^+ and $(H_3C_2O_2)^-$, and dried slowly at room temperature. XRD analysis of the phases and crystallographic structures of the samples was carried out using a Siemens D5000 X-ray diffractometer equipped with graphite monochromatized high-intensity $CuK_\lambda (\lambda=1.54178 \text{ Å})$. The Bragg angle 2θ ranges from 5° to 80° at a scanning rate of 0.06°/s. A scanning electron microscopy JEOL JSM 6400 was employed to investigate the surface morphology of the samples, due the ZnO is a semiconductor material is was necessary to cover them with a copper layer. A Tecnai F20 microscope operating at 200 kV with a field emission gun and EDS detector was used to further analyze the structural characteristics of the as-crystallized samples. The samples for HRTEM analysis were prepared by spreading a droplet of solution of ZnO particles onto a carbon film supported by a Cu grid and subsequent drying in vacuum.

RESULTS AND DISCUSSION

Figure 1, depicts the XRD data of the phases of as-crystallized samples and their crystallographic orientation synthesized with different molar ratio obtained by chemical precipitation. All of them had (00n) Bragg reflections attributed to the crystals structures and these samples show a high crystallization having strong Bragg reflections. The positions of the XRD peaks and the relative intensity of the all samples perfectly match the standard PDF values (JCPDS # 03-065-3411). In addition, the lattice constants of the as-crystallized array are measured to be a=0.3249 nm and c=0.5212 nm, which are in good agreement with the values of the wurtzite (ZnO) of the space group P63mc. The reflections before $25^0(2\theta)$ is due to the presence of the small amount of $Zn(OH)_2$ particles with wulfingite structure with lattice parameters of a = 0.319 nm and c = 0.465 nm with a space group of P3m1.

Yamabi [12] reported that ZnO has a value much less negative of ΔG° that $Zn(OH)_2$, in a pH strong alkali (9 to 14) and concentration of 1 M to1 μM, for that, we would expect to obtain a clear presence of crystallographic species of $Zn(OH)_2$ by presenting more ΔG°.

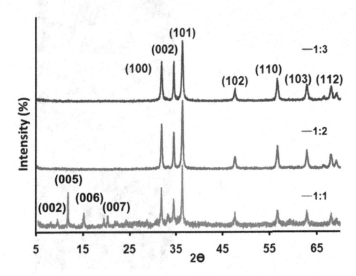

Figure 1. XRD patterns of ZnO and $Zn(OH)_2$ crystals synthesized a different molar ratio (The peak indexes are specified above the peaks).

Figure 2, present the EDS analysis and SEM image showing the morphology and chemical composition of the samples synthesized with molar ratio of 1:1, 1:2, and 1:3. It is possible to observe that the chemical composition is the same, the chemical composition is 37% for Zn and 63% for O, for the all molar ratio particles synthesized (Figure 2a, 2b and 2c). Only, oxygen and zinc signals can be attributed to the sample. The carbon and cupper peaks in the spectrum should arise from the background from the SEM grid and the particles metalized respectively. Figures 2 (d, e and f) shows the morphology of the particles obtained from the synthesis. It is well known that the crystallographic orientation and faceted structure of crystal samples depend on the supply of the sources, the adsorption of the reactants onto the growth sites, and the diffusion of the adsorbates on the substrate surface [8]. A hexagonal bar shape morphology particles were formed with molar ratio 1:1, the typical diameter is about 1 μm and the height is in the range of 0.7 to 1.3 μm which can be observing in figure 2d. Figures 2e and 2f synthesized with molar ratio of 1:2 and 1:3 respectively show different morphology and sizes showing circular shape morphology. The smallest size particles were obtained when the samples were synthesized with a molar ratio 1:3 which were less a 1μm of size. In the three cases the crystalline phase formed was wurtzite as shown in the XRD results (Figure 1).

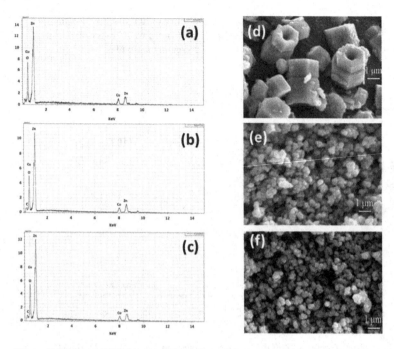

Figure 2. EDS and SEM images of the ZnO particles synthesized a different molar ratio. a) and d) EDS and SEM with molar ratio of 1:1, b) and e) EDS and SEM with molar ratio of 1:2, c) and f) EDS and SEM with molar ratio of 1:3.

The morphology and crystal orientation of the ZnO arrays were further investigated by using TEM and HRTEM. Figure 3, shows a bright-field TEM and HRTEM image top view of ZnO synthesized at different molar ratio. Figure 3a, 3b, and 3c show a TEM image of the particles synthesized 1:1, 1:2 and 1:3 respectively and Figure 3a', 3b' and 3c' the corresponding HRTEM and FFT results. It is possible to observe that the size was reduced when the stoichiometric changed from 1:1 to 1:3, all FFT image show a hexagonal structure which is agree with the results obtained by RXD analysis. The HRTEM image show that the crystals have an interatomic distance of 0.28 nm and 0.26 nm which corresponds to the (100) and (002) crystal plane of the ZnO crystal, the hexagonal FFT image near to the [1$\bar{1}$00] axis zone for the 3a' image.

Figure. 4a' and 4b' shows a poly-crystals FFT in the (001) and (101) planes respectively. Some adjacent nanocrystals can join together and grow to form a larger crystal. This suggests that 2D growth is dominant at this area and this stage of growth. Hence, keeping a low density of the nanocrystals is necessary for the formation of the crystals. On the other hand, for stoichiometric ZnO, the growth in the [0001] direction usually overwhelms other directions because the (0001) atomic planes are closed-packed with low surface energy [9]. The faceted structures observed in Figure 3a demonstrate that the growth velocity in lateral directions to that of c-axes do not decrease, while it is the same at the bases. Nonetheless, as (0001) planes are always the fastest growing crystallographic planes, they envelop the others and determine the preferable orientation to be [0001].

Figure 4 (a) shows HRTEM micrograph of the sample with a molar ratio of 1:1, which clearly have the definition of nanocrystals with hexagonal structure with the FFT image in the (002) plane. In Figure 4b we present the image of the filtered image showing the formation of nanocrystals with size less than 10 nm with interplanar distances of 0.235 nm which correspond to (002) plane assignable to $Zn(OH)_2$ crystals.

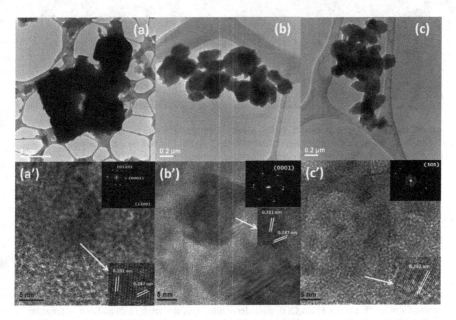

Figure 3. TEM and HRTEM images of the ZnO particles synthesized with different molar ratio a) and a') 1:1, b) and b') 1:2, c) and c') 1:3.

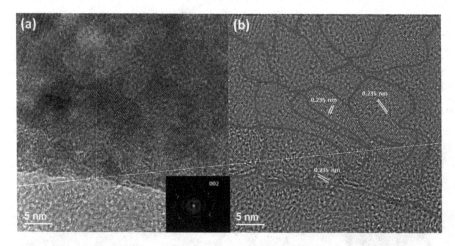

Figure 4. HRTEM images of the β-Zn(OH)$_2$ particles synthesized with molar ratio of 1:1a) HRTEM and FFT images showing the (002) orientation and b) Filtered image showing a nanocrystals with a interplanar distance of 0.235 that correspond to the (002) orientation.

CONCLUSIONS

In the present work, ZnO and Zn(OH)$_2$ crystals were synthesized by chemical precipitation process and have been characterized by XRD, SEM, EDS, TEM and HRTEM. A facile route for the synthesis of ZnO and Zn(OH)$_2$ crystals has been produced and reported. The method offers a very simple and low cost route for the production of ZnO and Zn(OH)$_2$ arrays. The presence of the limit reactant (NaOH 1:1) was critical for the formation of hexagonal shape morphology crystals of ZnO with a cover of Zn(OH)$_2$ particles while stoichiometric (1:2) and excess reagent (1:3) formed a circular shape of ZnO crystals without the presence of Zn(OH)$_2$ arrays. According to FFT analyses, the ZnO crystals had wurtzite orientation while the Zn(OH)$_2$ crystals had β-Zn(OH)$_2$ orientation. The crystal morphology individual exhibits uniform with preferred directions [0001] growth. The nuclei and subsequent development on surfaces (0001) on the crystals were observed. To the best of our knowledge, this is the first report of preparation of ZnO and Zn(OH)$_2$ arrays. The method can also be extended to the preparation of other materials.

REFERENCES

1. Q. Zhang, T. P. Chou, B. Russo, S. A. Jenekhe, G. Angew. Chem., Int. Ed. 47, 2402 (2008).
2. J. B. Baxter, A. M. Walker, K. van Ommering, E. S. Aydil, *Nanotechnology*, **17**, S304 (2006).
3. B. Xiang, P. Wang, X. Zhang, S. A. Dayeh, D. P. R. Aplin, C. Soci, D. Yu, D. Wang, *Nano Lett.*, **7**, 323 (2007).
4. Z. L. Wang, *ACS Nano*, **2**, 1987 (2008).
5. Q. Wan, Q. H. Li, Y. J. Chen, T. H. Wang, X. L. He, J. P. Li, C. L. Lin, *Appl. Phys. Lett.* **84**, 3654 (2004).
6. C. C. Li, Z. F. Du, L. M. Li, H. C. Yu, Q. Wan, T. H. Wang, *Appl. Phys. Lett.* **91**, 032101 (2007)
7. Y. F. Lin, W. B. Jian, *Nano Lett.* **8**, 3146 (2008).
8. J. Zhou, P. Fei, Y. Gao, Y. Gu, J. Liu, G. Bao, Z. L. Wang, *Nano Lett.* **8**, 2725 (2008).
9. C. Hanley, J. Layne1, A. Punnoose, K. M. Reddy, I. Coombs, A. Coombs, K. Feris, D. Wingett, Nanotech. 19, 295 (2008).
10. A. Khorsand, R. Razali, W. H. A. Majid, M. Darroudi, Int. J. Nanomed. 6, 1399 (2011).
11. L. Wang and M. Muhammed. J. of Mat. Chem. 9, 2871 (1999).
12. S. Yamabi, H. Imai, *J. of Mater. Chem.* **12**, 3773 (2002).

AUTHOR INDEX

SUBJECT INDEX

Al, 29, 83
amorphous, 37, 113

Bi, 3

Ca, 3
casting, 83
ceramic, 89
chemical synthesis, 29, 127
corrosion, 71
crystal growth, 45
crystalline, 19

diffusion, 105

film, 113
foam, 83
fracture, 71

geologic, 37
grain size, 89

hardness, 55

infrared (IR) spectroscopy, 45

kinetics, 105

Li, 29
luminescence, 113

Mg, 3
microstructure, 37, 55

nanostructure, 19, 119, 127

optical metallography, 63

radiation effects, 89

scanning electron microscopy
   (SEM), 11, 45, 63, 127
second phases, 55
steel, 63, 71, 97, 119
strength, 97
structural, 11

welding, 97, 119

x-ray diffraction (XRD), 11, 19, 105

Printed in the United States
by Baker & Taylor Publisher Services